Lecture Notes in Electrical Engineering

Volume 759

The book series *Lecture Notes in Electrical Engineering* (LNEE) publishes the latest developments in Electrical Engineering - quickly, informally and in high quality. While original research reported in proceedings and monographs has traditionally formed the core of LNEE, we also encourage authors to submit books devoted to supporting student education and professional training in the various fields and applications areas of electrical engineering. The series cover classical and emerging topics concerning:

- Communication Engineering, Information Theory and Networks
- Electronics Engineering and Microelectronics
- Signal, Image and Speech Processing
- Wireless and Mobile Communication
- Circuits and Systems
- Energy Systems, Power Electronics and Electrical Machines
- Electro-optical Engineering
- Instrumentation Engineering
- Avionics Engineering
- Control Systems
- Internet-of-Things and Cybersecurity
- Biomedical Devices, MEMS and NEMS

For general information about this book series, comments or suggestions, please contact leontina.dicecco@springer.com.

To submit a proposal or request further information, please contact the Publishing Editor in your country:

China

Jasmine Dou, Editor (jasmine.dou@springer.com)

India, Japan, Rest of Asia

Swati Meherishi, Editorial Director (Swati.Meherishi@springer.com)

Southeast Asia, Australia, New Zealand

Ramesh Nath Premnath, Editor (ramesh.premnath@springernature.com)

USA, Canada:

Michael Luby, Senior Editor (michael.luby@springer.com)

All other Countries:

Leontina Di Cecco, Senior Editor (leontina.dicecco@springer.com)

** This series is indexed by EI Compendex and Scopus databases. **

More information about this series at http://www.springer.com/series/7818

Njabulo Kambule · Nnamdi Nwulu

The Deployment of Prepaid Electricity Meters in Sub-Saharan Africa

Riding the Fourth Industrial Wave

 Springer

Njabulo Kambule
Department of Geography, Environmental
Management and Energy Studies
University of Johannesburg
Johannesburg, South Africa

Nnamdi Nwulu (ID)
Department of Electrical and Electronic
Engineering Science
University of Johannesburg
Johannesburg, South Africa

ISSN 1876-1100   ISSN 1876-1119 (electronic)
Lecture Notes in Electrical Engineering
ISBN 978-3-030-71219-8   ISBN 978-3-030-71217-4 (eBook)
https://doi.org/10.1007/978-3-030-71217-4

This Springer imprint is published by the registered company Springer Nature Switzerland AG
The registered company address is: Gewerbestrasse 11, 6330 Cham, Switzerland

Contents

Acronyms

AMI	Advanced Metering Infrastructure
AMP	Ampere
ANC	African National Congress
C	Cent
c/kWh	Cent per kilowatt hour
CESS	Collective Energy Switching Scheme
CFL	Compact Fluorescent Lamp
CIU	Customer Interface Unit
DEA	Department of Environmental Affairs
DME	Department of Minerals and Energy
DoE	Department of Energy
DSM	Demand Side Management
E	East
ED	Electric dispensers
EDM	Electricidade de Moçambique
EPBS	Electricity Prepaid Billing System
EPC	Energy Performance Certificates
Eskom	Electricity Supplying Commission
FBE	Free Basic Electricity
GW	Giga Watt
IBT	Inclined Block Tariff
INEP	Integrated National Electrification Programme
Kv	Kilo volt
Kw	Kilowatt
kWh	Kilowatt hour
LED	Light emitting-diode
M	Meter
N	North
NEP	National Electrification Programme
NERSA	National Electricity Regulator of South Africa
NTLs	Non-Technical Losses
PEH	Prepaid Energy Hub

PHCN	Power Holding Corporation of Nigeria
PURC	Public Utilities Regulatory Commission
RSA	Republic of South Africa
S	South
SABC	South African Broadcasting Commission
SECC	Soweto Electricity Crisis Committee
SMS	Short Message Service
SOWETO	South Western Townships
StatsSA	Statistics South Africa
ToU	Time of Use
TRA	Tembisa Residents Association
TV	Television
UBPL	Upper-Bound Poverty Line
UK	United Kingdom
UNIDO	United Nations Industrial Development Organisation
USA	United States of America
USD	United States Dollar
VAT	Value Added Tax
W	West
WB	World Back
WEC	World Energy council
WEF	World Economic Forum

Chapter 1
Introduction: Electrical-Energy Revolutions and Smart Grids

Energy is as ancient as time. All phenomena, historical inventions, industrialisation, civilisation, and modernisation unfolding in time, hinge on energy. It is an essential property that exists in natural material and systems. Through energy, atoms and molecules are set in random motion, enabled to interact and vibrate. It is thus an important mainstay of livelihood, matter, and space. It is also the cause and effect, as well as the measure of the historic, present, futuristic change in society [26, 45]. Energy continues to frame and drive development across global societies and as such has been acknowledged the currency of motion, development, and progress over time [2, 10, 11, 18, 23, 34, 39, 46, 47]. Hitherto, the role of energy in societal or human development remains uncontested. It has become a universally applied measuring standard for regional development. Improved access to energy has the propensity to translate in ameliorated standards of living [11, 45].

Different types of energy carriers have been discovered in the past waves of the revolution. Figure 1.1 is a historic overview and evolution of human reliance on energy [10, 11, 23, 26, 39, 44]. Historically, particularly during the Pre-Industrial era energy needs for human survival remained modest. The sun was the primary source of heat. With time, reliance shifted towards burned wood, dried dung, and straw for heat [47]. Sailing ships were the means of transportation through water and wind. Animals were fundamental instruments of intensive labour [2, 46].

Whereas access to a specific form of energy, i.e. electrical energy, has now enabled a more convenient livelihood for humanity, a significant portion of rural households in Sub-Saharan Africa still has limited access to it [2, 34, 45]. This in-turn has subdued development, and in an unfolding era of the fourth industrial revolution it is imperative to draw-out implications associated with this.

With this backdrop, the chapter commences with a review and discussion of the Industrial revolution, highlighting the different eras or waves that characterise it. Essentially we show that energy and electricity are at the centre of this phenomenon.

© The Author(s), under exclusive license to Springer Nature Switzerland AG 2021
N. Kambule and N. Nwulu, *The Deployment of Prepaid Electricity Meters in Sub-Saharan Africa*, Lecture Notes in Electrical Engineering 759,
https://doi.org/10.1007/978-3-030-71217-4_1

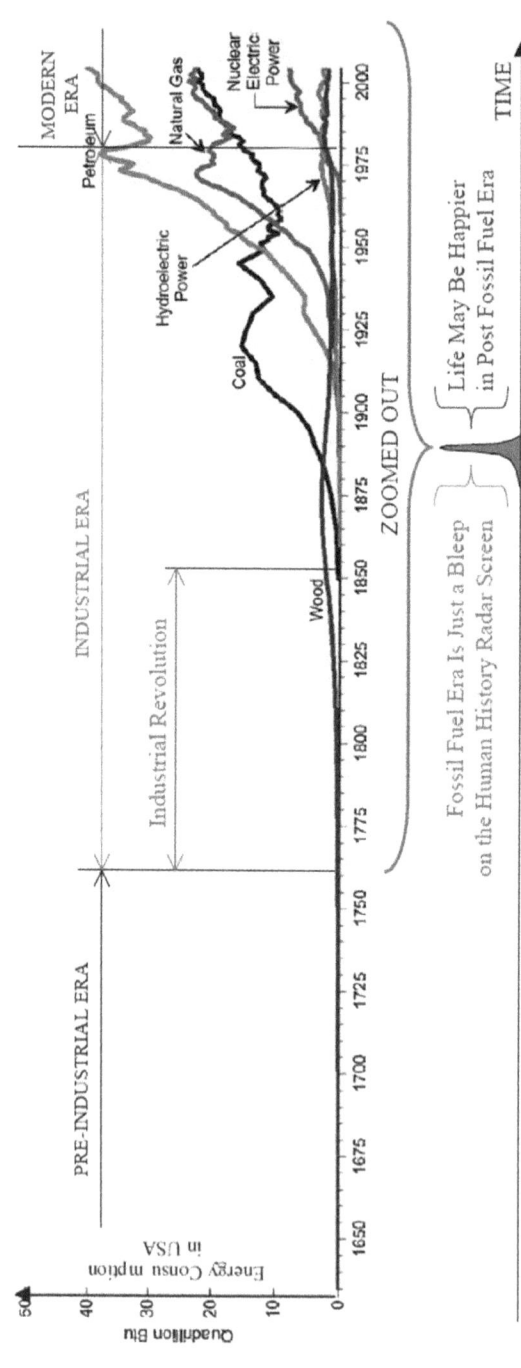

Fig. 1.1 Energy evolution [18]

We lastly focus on rural electrification and smart grid development, focussing on how mini-grid systems and smart prepaid electricity meters (SPEM) are fundamental building blocks for Sub-Saharan Africa's move and intention of fortifying its agenda on the fourth industrial revolution.

1.1 Industrial Electrical-Energy Revolutions Era

The Industrial Revolution unfolded between the eighteenth and nineteenth centuries. This period was characterised by a sudden surge in fossil fuel (i.e. coal, oil, and natural gas) based energy [2, 47]. The abundance of these resources and their use in daily human life radically transfigured the general societal livelihood [11]. Until recently, the era was generally divided into three waves (Figs. 1.1 and 1.2).

1.1.1 Era 1: Steam-Based Energy

The first wave of the 1700s Industrial Revolution mainly features inventions of steam engines by Thomas Newcomen, Savory, and James Watt, between 1712 and 1765, as a pillar that set the tone for intensive and unprecedented societal development through fossil fuels [11, 26, 39, 45, 47]. Locomotives and farm implements were powered by steam, and production for the first time was mechanised. On the other hand, coal became a useful resource for warming-up buildings and steel production. To date, more than 85% of the total share in modern energy use is from fossil fuels [19, 47].

1.1.2 Era 2: Industrial Electric-Energy

The invention of the first electric-generator utilising coal in the 1800s signalled the start of second and move to the third wave of the Industrial Revolution [11, 45]. This specific creation was shortly followed by Thomas Edison's electric light invention, which pro roved to be a watershed moment in the proliferation of electrical-energy. In the beginning, electricity use was specifically for telegraph and lighting applications [2, 4]. The invention of an electric system with a source of power resulted

Fig. 1.2 Waves of Industrial revolution [2, 11, 26, 39]

in Manhattan being the first region where mass electrification took-off [2, 11, 16, 26]. The availability of alternating current (AC) in the mid-1880s permitted for electricity to be transmitted and distributed over long distances. This was primarily because voltage could be increased (stepped-up) and decreased (stepped-down) [11, 18, 45]. At this point, electricity had become a market commodity with a price-tag attached. However, it was still available in 'local pockets' because "each utility, supplied electricity to its local city or neighbourhood and was completely isolated from others" with "each utility vertically integrated[1]" [4]. Electricity was more of a luxury than a necessity. It was expensive—statistically almost 70 times higher than the current average cost [2, 17, 18].

1.1.3   Era 3 and 4: Modern Clean Electrical-Energy

As opposed to the second wave—highly dependent on fossil fuels for survival—the third and fourth wave (twentieth and twenty-first century) marked the convergence of clean energy and collaborative communication [20, 25, 34, 39]. The Third wave of Industrial Revolution (starting in the 1960s) witnessed the use of digital computers and the internet resulted in the automation of manufacturing and disruption of the electrical-energy market [11]. The burgeoning market demand for electricity gadgets (telephone, television, and radio) meant that: electricity capacity and supply had to increase; and, the global economy was becoming electricity oriented (Fig. 1.2) [2].

Today, electricity has become a central feature of modernisation and development. Lloyd [19] and Vllasaliu [40] recently assert that, electricity "is such a necessary element in development that it should be seen as a basic right". Electricity accessibility or electrification has therefore been embraced as a socio-economic development project that different governments across the globe are committed to [7, 20, 25].

The youngest phenomenon is that of the fourth industrial revolution, wherein focus in the twenty-first century has transitioned towards new technologies that fuse the physical, digital, and biological realms [2, 7, 11] describes it as.

> …the era of linking the innovated existing first industrial revolution devices with the inno-
> vated existing second industrial revolution devices and innovated third revolution devices.
> The linking will be done with effective communication through programming.

It is globally recognised as the most important economic and societal trend as it affects all fields of human livelihood and will essentially transform the nature and ability of functionality of business and society in the next decades [7]. As the fourth wave continues to drive current and future markets, as well as human behaviour, one of the fundamental questions that policymakers, researchers, and academics need to address, is that of the impact of the wave on the electric-energy landscape

---

[1]Utility responsible for electricity generation, transmission, distribution to customers and subsequent billing.

in Sub-Saharan Africa. How can the wave be effectively used to improve household electrification? Can the wave be used to improve *connectivity* especially for households located in remote areas to participate in the mainstream *virtual* (online) economy? How will the regulatory landscape respond to this inevitable societal and economic transformation?

Electricity is at the centre of this revolution. According to [45], the wave will induce disruptions in three forms:

- Electrification
- Decentralisation
- Digitisation

These forms exist in a cyclic and interdependent manner. In electric-energy (or electrification) terms this wave has altered the process of electricity generation. The impact of burning fossil fuel for electricity generation has resulted in carbon-intensive based global economies [11, 20, 39, 47]. Consequently, attention in the fourth wave has increasingly been on the transition towards emerging clean electrical-energy and energy efficient technologies. While the fossil fuels used for electricity generation remain inexpensive, due to their abundance, but the environmental cost remains high. Nuclear energy, albeit being clean, is a contentious energy source in the modern era [25, 34, 45]. However, renewable energy sources are arguably universally accepted as mainstream instruments for global economic development. The energy sources include the sun, wind, biomass (e.g. garbage, and forest, and agricultural waste), geothermal heat extracted underground, ocean tides and waves, and small hydro [2, 26, 39]. Notwithstanding, geographical uneven distribution and intermittence of the noted renewable energy sources and smart grid software systems form part of the essence of the fourth wave [34]. In the light of the aforementioned global electric-energy revolutions, realistically, a majority of nations—a majority located in underdeveloped or developing regions—are struggling to evolve even beyond the second wave [25, 26]. There are of course several underlying reasons why this has been the case and the next section specifically discusses the role of improved of electrification in Sub-Saharan Africa as a mechanism of development in the fourth industrial revolution wave [2, 7, 11, 20, 45, 47].

1.2 Household Electricity in Sub-Saharan Africa

Improving electricity access accelerates human development [18–23]. The United Nations Sustainable Development Agenda 2030 aims at providing universal access to electricity by 2030. Improved electrification has the propensity of improving education and health services, as well as creating employment [39, 41]. Currently, the Sub-Saharan African region has an overall operational electricity capacity of about 100 GW. Of this total, 49% is from coal, 20% large hydropower, and 16% from gas [2, 20]. Poor infrastructural maintenance, management, and planning are among several factors that lead to relatively higher and unaffordable electricity generation and distribution costs in the region [19, 23, 25, 41].

Albeit the reality of uncertainty about the household electrification rate in developing countries, and Sub-Saharan Africa specifically, it is broadly accepted that close to 1.1 billion (14%) people worldwide still lack electricity [43]. More than 2.8 billion households rely solely on traditional biomass for heating and cooking and [42]. The low connection rates are linked to the high electricity connection charges, which entail geographical hurdles in the construction of transmission and distribution networks, and excessive stringency of technical standards, inefficient procurement practices [2, 6]. There is a clear need for governments to tailor individual development agenda based on electrification reality in the region.

Figure 1.3 is an overview of the household electrification rate progress in the Sub-Saharan Africa region between 2000 and 2017 [13]. While there is clear and positive progress in connecting households to the electricity grid, the connection rates in 2016 were still among the lowest (42%) in the world. Additionally, the connection rate is not homogenous; it is different across the region. For instance, in countries like Gabon, Kenya, and, Swaziland, the rate has been more rapid (rising by more than 50% between 2000 and 2016) than in countries like Zimbabwe, where connection rates have declined [39]. Figure 1.4, is an example of how the overall improved electricity accessibility in the region is characterised by heterogeneity [13]. This is an indication that electricity is an indicator of wealth disparity that exists within the region [23, 25, 41].

The International Energy Agency [13] reports that close to 95% of those without electricity are based between Sub-Saharan Africa and Asia [13]. In absolute and specific terms, more than 600 million (57%) (2 in every 3 people) of the population are without access to electricity in Sub-Saharan Africa. About 15 countries in that region have less than 25% electrification rates. These rates are possibly extremely lower with the exclusion of countries like South Africa which is responsible for the majority (90%) of household connection rates in the region. In a recent survey of 22 Sub-Saharan Africa countries, it was found that "income levels…seem to be key determinants of electricity use" [37].

Relative to the rest of the world, household electricity consumption rates in Sub-Saharan Africa are significantly lower—especially in the exclusion of South Africa

Fig. 1.3 Electrification access rate for selected countries [13]

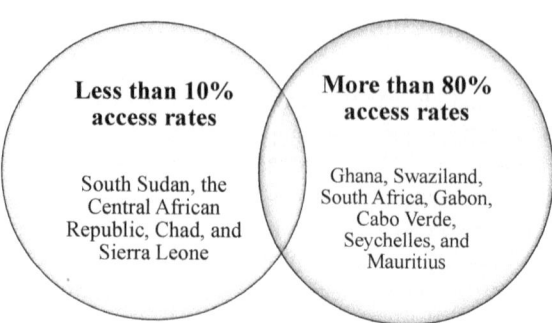

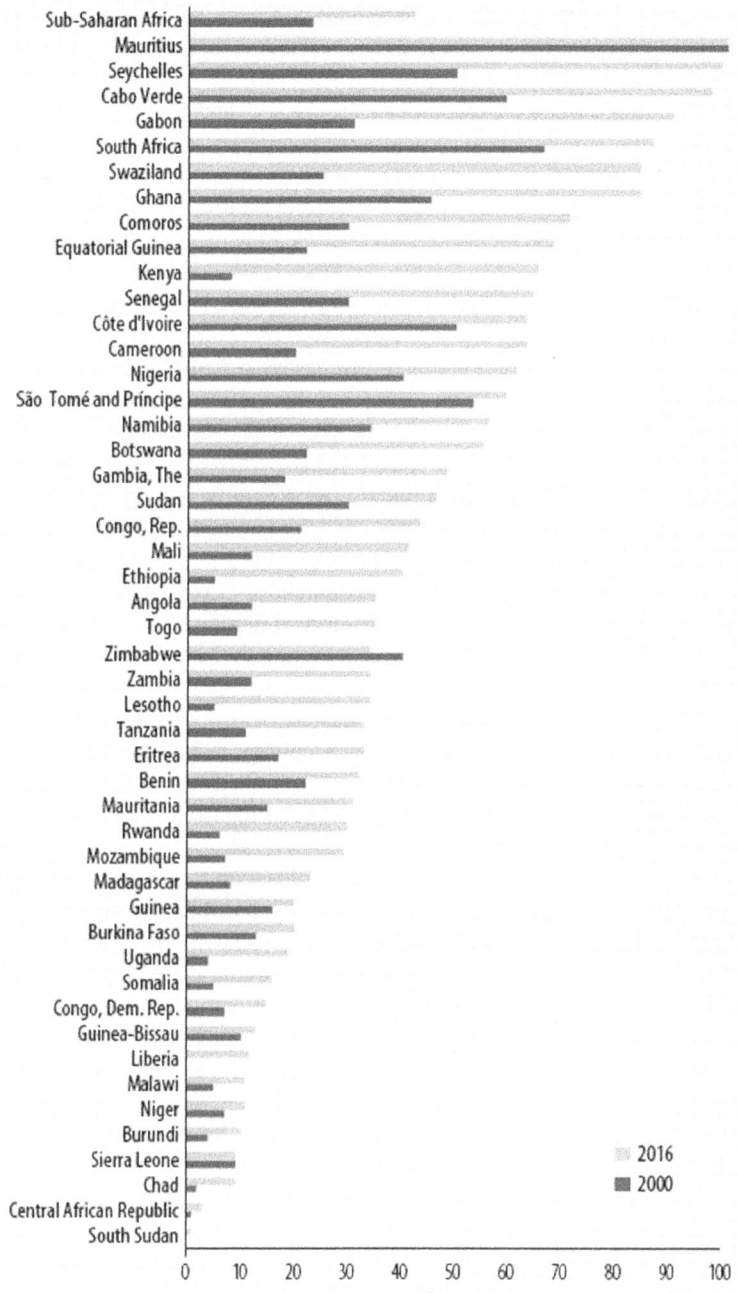

Fig. 1.4 Electrification access rate for Sub-Saharan Afrin Region [13]

[19, 23, 25, 41]. In comparison to countries like China and the United States of America (USA) with electricity consumption per capita ranging between 12 and 15%, the Sub-Saharan region has less than 3% [2]. The electricity tariffs in the region tend to be below production costs (except for two countries; they remain relatively high in global terms. Utilities continue to be unreliable guarantors of private investment [20, 41]. Again, the construction of new power plants/infrastructure is close to impossible and electricity shortage is entrenched [19, 37]. Regulators, therefore, opt towards increasing electricity tariffs for utilities to cover the full costs of the electricity supply service.

Faced with growing electricity demand, systems losses (i.e. non-technical losses—unaccounted electricity usage due to *inter alia* theft, utilities fail to correctly bill customers or technical losses along the transmission and distribution networks) are a reality in the Sub-Saharan region [19]. Thus far, only four countries (i.e. South Africa, Botswana, Mauritius, and Lesotho) have well managed their losses within the World Bank's 'Good Standard" threshold (i.e. loss of ≤10%). By contrast, eight countries (Central African Republic (CAR), Republic of Congo, Sap Tome and Principe, Comoros, Sierra Leone, Nigeria, Madagascar, and Cape Verde) experience losses of ≥30% [37]. In simple terms this means that for every 10 kWh generated by utility only 7 kWh can be financially accounted for [2, 19].

Until recently, household livelihood and survival in Sub-Saharan Africa has been fixated on fossil fuels [20, 23]. The region is among the best for solar photovoltaic (PV) and small-large scale hydropower. On the other hand, western Sub-Saharan Africa has large natural gas reserves and Eastern with high geothermal potential [19]. High investments in these sectors are necessary to harness optimal benefits from these resources. Hitherto, decentralised electrification in a form of renewable energy sources (such as solar home systems) is becoming the preferred alternative for investment among private companies [23, 25, 27, 41]. This is generally because rural households, constituting close to 80% of households without electricity, are sparsely populated, thus making the connection cost level higher than normal [2, 13, 19, 20, 39, 41] . Off-grid, mini-grid, and stand-alone renewable energy systems have thus gained traction in the region as instrumental in improving electricity accessibility in the region. Low-income households may access low electrical appliances (e.g. light-bulb and cellphone), and middle to high-income earners may access low or medium power electric appliances (i.e. television, refrigerator, air, and cooling fan) [19].

Schwab [35] has argued that "Many (Sub-Saharan African states) have still not enjoyed the benefits of the second industrial revolution". Sub-Saharan Africa governments need to create conducive investment environments and be proactive in adopting emerging household technologies. Smart household technologies may elicit positive spin-offs for the region [1, 23]. For example, the development and introduction of smart grid systems may assist improve the efficiency of distributing electricity across several homes in very remote locations [19, 20, 25, 27, 41].

1.3 Rural Electrification, Smart Grid Development, and Decentralisation

From the third wave into the fourth, several electrical technologies have had significant implications on the Sub-Saharan electricity grid. This is what is commonly attributed to as decentralisation. The phenomenon is comprised of the four following factors:

- *Distributed Generation*: This is electricity generated from renewable energy sources, particularly renewable energy sources. Sub-Saharan Africa is endowed with solar-based resource and countries like South Africa, Namibia, Kenya, and Ghana are identified as leading solar-based energy markets. Consequently, the price of residential solar PV has become cost-competitive as it has plummeted by more than 61% since 2012 [11].
- *Distributed Storage*: this is a mechanism of collecting electrical energy to utilise it during peak periods or for backup. This flattens demand peaks and valleys. With Sub-Saharan Africa again endowed with Hydro and Gas, this is another area that will fast have an impact on the overall regional grid.
- *Energy Efficiency*: the ability to use and reduce energy without compromising the quality of the final product and reducing demand is important for energy economies. The goal is to reduce electricity consumption to as reasonably low levels as possible, especially within household settings. According to [13], household energy consumption for lighting has decreased by about 80% as incandescent lamps have supplanted with compact fluorescents and Light Emitting Diode (LED) bulbs. By proactively applying energy efficiency measures, South Africa can mitigate its electricity demand by more than 16% by the year 2030 without restricting economic growth [1].
- *Demand Response*: enables the control of energy consumption at peak demand and high pricing periods by giving price or volume signals, and sometimes financial incentives to reduce demand at certain strategic periods of the day [11].

A plethora of challenges continue to hamper the distributed generation and grid connection project in Sub-Saharan Africa [23]. Energy needs, at least for lighting, are still based on the availability or usage of candles, kerosene lamps, generators, torchlights. Wood and dung are cooking the main energy sources for rural households [20]. The demand for electricity in these areas thus remains low. A socio-economic reason associated with is that a majority of the households relatively earn below-average income and struggle to afford commercially available (on-grid) electricity [1, 2, 23, 39, 41].

Besides this fact, the extension of existing grid networks to include rural areas also stands as an important hurdle for improved electrification in the region. A majority of rural households remain without electricity because of grid extension issues. There is a growing realisation that achieving universal electrification rates through expanding the traditional on-grid system is largely impossible [1, 39, 47]. This is due to issues of topography, low population density, as well as the fact that most rural communities are isolated in location,meaning that the cost of transmission and distributing electricity to the areas make may prove expensive and cumbersome [27]. The World Bank [44] found that the Sub-Saharan region has the highest number of countries with high connection charges (i.e. $\geq$\$100 per customer at the lowest connection service rating). The aforementioned are among several factors that have prolonged the delay of connecting households to the grid. According to [11]

> The fourth may finally deliver electricity because it no longer relies on centralised grid infrastructure...A smart grid can distribute power efficiently across a number of homes in very remote locations.

Authorities across the region are therefore growingly considering smart mini-grid grid systems as an alternative approach to leverage and bolster the electrification rate in the region [1, 20, 27, 36]. The application of renewable energy technology has become one of the most pragmatic approaches to improve rural electrification, especially in the context of geographically challenging areas [41, 47]. Renewable energy currently is the most preferable, sustainable, and viable energy source for electricity accessibility in the region. Governments and public utilities are growingly investing in this source of energy. Sub-Saharan Africa is endowed with a variety of renewable energy sources, including solar, wind, hydro, geothermal, and biomass [2, 23].

Among the various energy sources, solar energy has the largest potential to assist with improved electrification. About 10 TW of energy can be harnessed for solar power; wind about 1, 300 GW, and geothermal around 1 GW [9, 11, 39]. Solar mini-grid (decentralised) systems have also become popular, enabling an easier and cheaper means for electrification [1, 47].

The phenomenon of *digitisation* is imperative in the fourth wave of the industrial revolution. According to ROSCONGRESS [34], "digitalization is the driver of efficiency growth in energy". Digitalisation is the ability of digital technologies, located across the grid to communicate and give useful customer-based data for better operational and grid management [11]. With this, utilities have access to real-time operations of the grid network and its connected resources and may also collect network data to optimise situational awareness and reduce system non-technical losses. Smart meters are an example of digital technologies that assist utilities to provide services in remotely located areas.

Thus far, the average electricity grid access rate remains at about 20%, so issues of transmission and distribution prove to be a problem when it comes to connecting households to the grid. The application of min-grid systems has become a viable and cost-effective solution for electrification challenges (e.g. poor reliability and inconsistent quality) in most countries (rural areas) in the region. Sub-Saharan Africa has close to 2000 mini-grid sites, of which about 40% of those are solar projects [12]. Rural areas in countries like South Africa, Gambia, Botswana, Kenya, and Tanzania have operational solar mini-grid systems. Some of the systems are solar-battery hybrid systems [23, 47].

In Nigeria, only about 36% of the country's electrified households are based on a centralised system; about R20 billion is being invested in mini-grid systems to improve electrification [12]. The use of these systems is expected to grow exponentially (16, 000 sites) in the region by 2023. Household mini-grid systems are currently normally characterised by a fee for service, rely on dispersed technology (e.g. solar heater systems), fixed monthly payments, use of batteries, and prepaid meter cards [1, 3, 23, 36].

At operations level, in the context of smart household prepaid electricity meters, the growth of renewable energy mini-grid systems has implications on the metering, billing, and collections systems. The installation of smart prepaid metering systems may enable utilities to accordingly measure electricity demand and usage [1, 20, 23, 27, 28, 36]. With that, operators or regulators are enabled to determine and control household electricity consumption fees or tariffs [41]. As compared to a normal payment meter, a smart prepaid electricity meter bears the following features and advantages [1, 3, 39, 41, 47]:

- Quantity of electricity used
- Time of use
- Appliances associated with each connection
- Improved operational efficiency
- Reduced operations and maintenance costs.

As it will be seen later in the book, smart prepaid meters are becoming popular models of payment in Sub-Saharan Africa [1, 27, 36, 47]. In a market where populations have access to smart mobile phones, smart prepaid electricity meters provide a smart option of purchasing electricity using mobile phones. So, instead of a reliance on vouchers or scratch card methods mobile phones are increasingly offering an improved and more reliable payment option as well as revenue collection [2, 3, 41]. This means that households with access to the network may purchase electricity at any point of the day and thereby improving sales. This is the advantage that comes with the Fourth Industrial Revolution whereby customers or prosumers can have direct control and impact on the final bill and be accountable for the energy choices made.

In a study conducted in Nigeria, the revenue collection rate of 9 audited mini-grid projects with smart prepaid electricity meters installed ranged between 98–100% [23, 27, 39]. From these projects it was found that successful mini-grid projects ensure about 15−20% returns. This is however reliant on a variety of factors which include site selection, community engagement, demand, stimulation, ownership, and regulatory support [1, 3, 28]. Table 1.1 is a reflection potential links between each of the noted factors to the effectiveness of a smart prepaid electricity meter project.

Under a cloud that is now driven by *virtual connectivity* smart or mini-grid technologies are important agents of driving this connection for remotely located regions (i.e. rural areas) [34]. Cost-effective mini-grid solar system technologies can be used for a variety of reasons including charging laptops and cellphones to connect with

Table 1.1 Mini-grid related factors and implications associated with the use of prepaid electricity meters [1, 2, 20, 23, 27, 36, 47]

Factors	Implication on prepaid metered mini-grid system
Site selection for economic viability	For whichever site selected it is important to first determine the economic viability of the prepaid electricity meter programme. In this conducting a socio-economic or situational impact assessment study becomes necessary. This may, for instance, assist to gauge the extent of use of smart mobile phones and therefore improve the convenience of the technology or programme. Again, an assessment of this nature will reflect the willingness of consumers to use and pay for prepaid meter based electricity. The latter will help ultimately present minimal concerns about safety or social unrest
Community involvement	The involvement of the community within which the prepaid electricity meter project will be carried out is cardinal. The community may be comprised of various stakeholders, i.e. local leaders, government officials, civil organization leaders, and general community members. The process of involvement has to be from inception or conception phase and beyond. This may be undertaken through regular public meetings. Such an engagement will ensure local buy-in and the sustainability of the prepaid meter programme within the mini-grid project. Community's involvement will also ensure that all stakeholders are invested in the project and therefore households, in particular, are likelier to embrace the value of using the technology, which in turn increases the physical and financial security of the project
On-going community engagement	Community involvement is not to be a once-off event, but an on-going process. The purpose of this is to maintain satisfaction and constant identification of emerging operational issues. In specific terms, on-going engagement with include inter alia frequent site visits, regular meetings with local community stakeholders. There are several advantages attached to this: • There is growing trust between authorities (government or utilities) and the community • Burgeoning or maintained customer/household willingness to use prepaid electricity meter technology • Programme remains relevant to the prevalent household socio-economic conditions • Improved economic viability

(continued)

Table 1.1 (continued)

Factors	Implication on prepaid metered mini-grid system
Demand Side Management (DSM)	Prepaid electricity meters work most effectively within DSM initiatives such as energy efficiency. The installation of stand-alone solar home systems and prepaid electricity meters should also be accompanied by DSM initiatives. In a mini-grid system DSM may assist in the balancing of load profiles in the community. The provision of energy efficient incentives through the use of affordable Light Emitting Diode (LED) lighting, refrigeration, heating, and cooling systems, will improve the socio-economic efficacy of the prepaid meter programme in the project site
Capacity building	Project managers need to invest time and resources in capacity building. This may come in a form of workshops that seek to educate households on the benefits of energy efficient appliances. These workshops will have to be held periodically and in areas closest to the mini-grid sites selected
Formulation of regulations	The existence of policy to govern mini-grid projects is not only pivotal for sustainable investment purposes but also to protect the consumer/households. Poor regulatory governance may result in consumers being socio-economically marginalised by prepaid meter technology. So, the presence and carrying out of mini-grid regulations have to be tandem with that of formulating and constant updating of prepaid electricity meter regulations, to protect energy vulnerable households
Energy poverty	High and escalating electricity tariffs have the propensity of entrenching energy poverty. Although the Sub-Saharan Africa electricity market continues to generally possess relatively low tariffs, their increase over time in selected countries such as South Africa have raised issues of affordability and widened socio-economic inequality. Improved education on energy efficiency and deployment of the requisite technology may assist in protecting vulnerable energy

the rest of the world for business or school purposes. This is becoming an integral economic opportunity and a component of social development that government to increasingly embrace and extensively invest on (i.e. human and monetary capital) [11, 20, 23, 27, 36, 41, 47].

Given the foregoing context, this book aims is to generate specific knowledge on the deployment of prepaid electricity meters in Sub-Saharan Africa. Thus far, in the past three decades there have been but only a handful of scholars that have endeavoured to research about this market [3, 5, 13–16, 21, 22, 24, 33, 36, 38]. The research, although beneficial but it has only looked at country-specific contexts.

No work has attempted to focus on the impact of this technology across the Sub-Saharan Africa region. *Why is this necessary*? Recent research has shown, confirming studies by O'Sullivan et al. [29–32] that this technology has the propensity of further entrenching inequality and poverty. Brynjolfsson and McAfee [8] have additionally warned that the Fourth Industrial Revolution could engender deepened inequality. Much focus has thus far been on how utilities have managed to recover debt but [1] has argued that "smart electricity planning considers the immediate needs of people." So, the real task is also in devising a sustainable means of averting energy inequality through studying its impact on the populace.

To initiate and drive this inclusive growth, the book provides a Sub-Saharan Africa prepaid electricity meter oriented framework that will render the technology to be a socio-economic value-add in the region. The knowledge shared in the respective chapters is trans-regional and can, therefore, be of assistance in other developing, and perhaps also developed regions. The government, policy and decisionmakers, public utilities, and non-governmental organisations involved in energy matters will find this valuable as it adds to the scholarly debate on the effectiveness of the technology, particularly in the Fourth Industrial Revolution. The scientific knowledge given creates a possibility for the poor to also benefit from this unfolding phenomenon of the Fourth wave.

1.4 Structure

The book is comprised of 8 chapters that look into the development and status quo of the prepaid electricity meter market in the Sub-Saharan Africa region. The diagram below serves as a pictorial illustration of the structure, dynamics, and synchrony of ideas that will be presented in the respective chapters of the book.

Chapter 1 echoes the necessary context within which the discussion of Chaps. 2–8 leans. This chapter focuses on household electrical-energy use in the Fourth Industrial Revolution. We briefly track the evolution of the nature of energy and its consumption within the four ways of the revolution. We share perspectives on the location of Sub-Saharan Africa in the revolutions. While it is generally accepted that many countries in the region still lag behind and possibly trapped in the second wave [11, 35], we discuss prepaid electricity meters are an essential agent reform for improved electricity accessibility. Most importantly, we argue that smart prepaid electricity meters are a fundamental pillar and catalyst for the region to effectively ride the Fourth Industrial wave.

Chapter 2 starts by gives background specific information about prepaid meters as a technology device supplanting conventional meters. Building from Chap. 1, this chapter elaborates on how prepaid electricity meters have gained traction over the past 3 decades in the Sub-Saharan region. We proffer a historical narrative of the development of the prepaid electricity meter market in the region. Beyond this, we also provide statistics on the extent of device adoption and see how South Africa, Mozambique, and Tanzania are the current leading users of the technology.

Chapter 3 serves to uncover the fundamental motives predicating, not on the deployment but the surging deployment rate in the Sub-Saharan regions. Using the identified motivations, we begin to share perspectives that challenged mainstream ideas about the beneficial nature of the technological device among low-income household settings. Consequently, we build a framework model that we see as a conduit towards rendering the technology effective in this particular region. Of course, this is following our view that the current framework for technology rollout is not in touch with the poor household socio-economic realities, and therefore threatens the already vulnerable households of the region.

Chapter 4 delves into fundamental arguments and perspectives around the nexus between prepaid electricity meters and energy poverty. We ask, is there a link? If there is, what is the nature of the relationship? Answers to these questions lead us to conclusions that will be of assistance to governments, policy and programme designers, civil society, and public utilities, for decision making.

Chapters 5–7 compartmentalises the nature of technological effect within three sub-regions of Sub-Saharan Africa. Due to the scarcity of data we have calculatedly selected leading countries (with the deployment) with reliable data to represent our view and conclusions about the regions experience, location, and status quo with the technology:

• Western Africa (Ghana and Nigeria)
• Eastern Africa (Tanzania and Kenya)
• Southern Africa (South Africa and Mozambique).

The Northern region was omitted because of scarcity in data and research.

The last chapter focuses on offering summated points on the knowledge shared throughout the book, by giving a list of ten lessons learnt over the past three decades. It is our hope as authors that with this book being the first of its kind in the Sub-Saharan Africa continent, not that we know of any in any continent, better-informed decisions may be carried out by policymakers and public utilities as we ride the Fourth Industrial wave.

Book Outline: Chapter 1 - 8

The Industrial Revolution, Electrification, Smart Grid development

Chapter 1 serves to provide contextual detail predicating the central theme of the book which is the deployment of prepaid electricity meters in Sub-Saharan Africa. The chapter unfolds in the in view of the Fourth Industrial Revolution and its ability to improve connectivity in the region.

Chapter 2 explores the historical narrative of the prepaid electricity market by tracing its adoption surge by the Sub-Saharan market from the 1980s to date.

Rationale Part 1 Historical Development and Market expansion

Rationale Part 2: A Misdiagnosis of Non-Payment and Electricity Theft

Chapter 3 probes the deeper rationale behind the technology and expanding prepaid electricity market. The chapter asks, are policy and decision makers cognisant of the implications associated with such a surge?

Chapter 4 aims on showing the relationship between prepaid electricity meter market and energy poverty. A framework to minimise this reality is proposed to policy and decision makers of the region.

Prepaid electricity meters & energy poverty: Lessons from South Africa

Status & Challenges Prepaid Electricity Meters in selected Western African Countries

Chapter 5 shares knowledge on the experiences of the Western Sub-Saharan Africa region by noting the challenges and status quo of the prepaid electricity meter market among selected countries.

Chapter 6 provides knowledge on the experiences of the Eastern Sub-Saharan Africa region by noting the challenges and status quo of the prepaid electricity meter market among selected countries.

Status & Challenges Prepaid Electricity Meters in selected Eastern African countries

Status & Challenges Prepaid Electricity Meters in selected Southern African countries

Chapter 7 discusses the experiences of the Southern Sub-Saharan Africa region by noting the challenges and status quo of the prepaid electricity meter market among selected countries.

Chapter 8 gives guidance to regional authorities and policy makers on how the project of prepaid electricity meter markets should be framed, by considering the lessons learnt about the technology in the regional market.

Lessons Learnt & Policy Recommendations

1.5 Highlights

- Energy is an essential component of the four waves of the Industrial Revolution. While the Fourth wave of the revolution is driven by electrification, more than 600 million ($\approx$57%) people remain without electricity.
- Notwithstanding the pockets of electrification achievements in countries such as South Africa, a significant portion of Sub-Saharan Africa continues to feature within the second wave of the revolution and this is due to continued dependence on traditional sources of energy and poor access to electricity services.
- Smart grid and renewable energy systems are being recognised by regional authorities as a sustainable approach towards, not only improved electrification rates in Sub-Saharan Africa but also in establishing fundamental pillars for the Fourth Industrial Revolution.
- The rural household electrification project with solar mini-grid systems, and using smart prepaid electricity meter among technology is gaining popularity (2000 mini-grid sites) across the region because of its viability (in the context of challenges with grid extension) and its convenience in cost recovering for utilities.
- Smart mini-grid systems are a cost-effective way of ushering *virtual connectivity* in the Sub-Saharan region. This can assist drive economic and social change in most remote areas of the region.
- Authorities in the region should create conducive investment (e.g. improved governance systems and regulatory framework) milieu for further development of mini grid systems and the adoption of household-based smart grid technologies (e.g. mini-solar technology and smart prepaid electricity meters).

References

1. Abrahams Y, Fischer R, Martin B, McDaid, L (2013) Smart electricity planning. Project 90
2. African Development Bank (2019) Potential of the fourth Industrial revolution in Africa: study Report unlocking the potential of the fourth industrial revolution in Africa. Technopolis and Research ICT Africa and Tambourine Innovation Ventures, Oct 2019
3. Aliu IR (2020) Energy efficiency in postpaid-prepaid metered homes: analyzing effects of socio-economic, housing, and metering factors in Lagos. Energy efficiency, Nigeria. https://doi.org/10.1007/s12053-020-09850-y
4. Anderson R, Fuloria S (2011) On the security economics of electricity metering. https://www.econinfosec.org/archive/weis2010/papers/session5/weis2010_anderson_r.pdf (Accessed 8 Mar 2020)
5. Azila-Gbettor EM, Atatsi EY, Deynu F (2015) An exploratory study of effects of prepaid metering and energy related behaviour among Ghanaian household. Int J Sustain Energ Environ Res 4:8–21.
6. Blimpo M, McRae S, Steinbuks J (2017) Electricity access charges and tariff structure in Sub-Saharan Africa
7. Brookings (2020) Capturing the fourth industrial revolution. https://www.brookings.edu/blog/africa-in-focus/2020/01/10/a-national-strategy-for-harnestoessing-the-fourth-industrial-revolution-the-case-of-south-africa/ (Accessed 07 May 2007)

8. Brynjolfsson E, McAfee A (2014) The second machine age: work, progress, and prosperity in a time of brilliant technologies. W. W. Norton & Company, New York
9. Cambridge Africa (2018) Why Africa is the next renewables powerhouse. https://www.cambri dge-africa.cam.ac.uk/ (Accessed 08 Mar 2020)
10. Centre of Excellence in Financial Services (COEFS) (2020) The impact of the 4th industrial revolution on the South African financial services market. https://www.coefs.org.za
11. Chukwuemeka D, Nonso NS, Chukwunonso ES, Casmir R (2018) Electricity and the fourth industrial revolution: implications for the Nigerian economy
12. ESI Africa (2018). https://www.esi-africa.com/industry-sectors/generation/nigeria-scaling-mini-grids-could-stimulate-20bn-investment/ (Accessed 08 Mar 2020)
13. International Energy Agency (2019) Africa energy outlook 2019. https://www.webstore.iea.org/download/summary/2892?fileName=French%20Summary-Africa%20Energy%20Outlook%202019.pdf (Accessed 05 Mar 2020)
14. Kambule N, Yessoufou K, Nwulu N (2018) A review and identification of persistent and emerging prepaid electricity meter trends. Energy Sustain Dev 43:173–185
15. Kambule N, Yessoufou K, Nwulu N, Mbohwa C (2018) Temporal analysis of electricity consumption for in prepaid- metered low-and high-income households in Soweto, South Africa. Afr J Sci Technol Innov Dev. https://doi.org/10.1080/20421338.2018.1527983.
16. Kambule N, Yessoufou K, Nwulu N, Mbohwa C (2019) Exploring the driving factors of prepaid electricity meter rejection in the largest township of South Africa. Energy Policy 124:199–205
17. Kambule N (2019) An evaluation of prepaid electricity meters acceptance and efficacy among low-income households in Soweto, South Africa (Doctoral Thesis)
18. Kostic MM (2007) Energy: global and historical background. Encyclopaedia Energy Eng 1–15
19. Lloyd PJ (2018) The role of energy in development. J Energy South Afr 1:54–62
20. Lucas P, Dagnachew AG, Hof AF (2017) Towards universal electricity access in sub-saharan Africa: a quantitative analysis of technology and investment requirements. BL Netherlands Environmental Assessment Agency, The Hague
21. Makonese T, Kimemia DK, Annergarn HJ (2012) Assessment of free basic electricity and use of pre-paid meters in South Africa. http://conferences.ufs.ac.za/dl/Userfiles/Documents/00000/577_eng.pdf. Accessed 13 Oct 2016
22. Malama A, Mudenda P, Ng'ombe A, Makashini L, Abanda H (2014) The effects of the introduction of prepayment meters on the energy usage behaviour of different housing consumer groups in Kitwe, Zambia. AIMS Energy 3:237–259
23. Morrisey J (2013) The energy challenge in Sub-Saharan Africa: a guide for advocates and policy makers part 2: addressing energy poverty
24. Mwaura FM (2012) Adopting electricity prepayment billing system to reduce non-technical energy losses in Uganda: Lesson from Rwanda. Utility Policy 23:72–79
25. Nalubega T, Uwizeyimana DE (2019) 'Public sector monitoring and evaluation in the fourth industrial revolution: implications for Africa'. Afr Public Serv Delivery Perform Rev 1:18. https://doi.org/10.4102/apsdpr.v7i1.318
26. Ndung'u N, Signé L (2020) Capturing the fourth industrial revolution: a regional and national agenda. The fourth industrial revolution and digitization will transform Africa into a global powerhouse
27. Northeast Group (2014) Sub-Saharan Africa Electricity Metering: Market Forecast (2014 to 2024). Northeast Group, LLC.
28. Nugroho SB, Zusman R, Nakano K, Takahashi K, Koakutsu RL, Kaswanto N, Arifin A., Munandar HS, Arifin M, Muchtar K, Fujita T (2017) The effect of prepaid electricity system on household energy consumption—the case of Bogor, Indonesia. In: Urban transitions conference, pp 642–653
29. O'Sullivan KC, Howden-Chapman PL, Fougere G (2011) Making the connection: The relationship between fuel poverty, electricity disconnection, and prepayment metering. Energy Policy 39:733–741
30. O'Sullivan KC, Howden-Chapman PL, Fougere G (2015) Fuel poverty, policy, equity in New Zealand: The promise of prepayment metering. Energy Res Soc Sci 7:99–107

31. O'Sullivan KC, Howden-Chapman PL, Fougere GM, Hales S, Stanley J (2013) Empowered? Examining self-disconnection in a postal survey of electricity prepayment meter consumers in New Zealand. Energy Policy 52:277–287
32. O'Sullivan KC, Viggers HE, Howden-Chapman PL (2014) The influence of electricity prepayment meter use on house hold energy behaviour. Sustainable Cities Soc 13:182–191
33. Quayson-Dadzie J (2012) Customer perception and acceptability on the use of prepaid metering system in Accra West region of the electricity company of Ghana. http://dspace.knust.edu.gh/bitstream/123456789/4900/1/Quayson-Dadzie%20John.pdf. Accessed 16 Oct 2017
34. ROSCONGRESS (2020) The electric power industry: challenges of the fourth industrial revolution. https://www.roscongress.org/en/news/mirovaja-elektroenergetika-vyzovy-chetvertoj-pro myshlennoj-revoljutsii-/ (Access 07 May 2020)
35. Shwab (2018) The 'fourth industrial revolution': potential and risks for Africa. https://www.naci.org.za/nstiip/index.php/analytical-contributions/technologicalprogress/40-the-%27fourth-industrial-revolution%27-potential-and-risks-for-africa (Accessed 03 May 2020)
36. Smart Energy International (SEI) (2017) Analysis: prepaid electricity metering in Africa. https://www.smart-energy.com/features-analysis/analysis-prepaid-electricity-meters-africa/ (Accessed 13 May 2020)
37. Streatfeild JEJ (2018) Low electricity supply in sub-saharan Africa: causes, implications, and remedies. J Int Commer Econ. https://www.usitc.gov/journals (Accessed 05 Mar 2020)
38. Tewari D, Shah T (2003) An assessment of South African prepaid electricity experiment, lessons learned, and their policy implications for developing countries. Energy Policy 31:911–927
39. United Nation Industrial Development Organisation (UNIDO) (2017) Accelerating clean energy through Industry 4.0 manufacturing the next revolution. Inclusive and sustainable industrial development
40. Vllasaliu D (2015) Factors affecting the regular monthly payment of electricity bills in Hajvali. Rochester Institute of Technology, RIT scholar works. https://www.scholarworks.rit.edu/theses (Accessed 05 Mar 2020)
41. Wiese D (2007) Rural electrification in Africa. West African Power Industry Convention, Abuja, Nigeria
42. Winkler H, Simoes AF, Rovere EL, Alam M, Rahman A (2012) Access and affordability of electricity in developing countries. World Dev 6:1037–1050. https://doi.org/10.1016/j.worlddev.2010.02.021
43. Wolfram C, Shelef O, Gertler P (2012) How will energy demand develop in the developing world. J Econ Perspect 1:119–138
44. World Bank (2017) State of electricity access report. https://www.worldbank.org
45. World Economic Forum (WEF) (2017) The Future of electricity new technologies transforming the grid edge. REF 030317
46. World Bank (2018) Access to electricity (% of population). https://www.data:worldbank.org/indicator/EG.ELC.ACCS.ZS. (Accessed 10 Apr 2018)
47. Xu M, David JM, Ki HM (2018) The fourth industrial revolution: opportunities and challenges. Int J Fin Res 2:90–95

Part I
Smart Prepaid Electricity Meters

Chapter 2
Rationale Part I: Historical Development and Market Expansion

2.1 Introduction

Africa is part of the global community that appreciates the value of smart prepaid electricity meters (SPEM) [6]. From the first introduction of the technological piece in 1988 to date, the continent has witnessed a surge in the rate of adoption—the rate is set to reach 234% by 2034 [19]. Faced with this reality it is important to review the historical and current role of the technology in the continent, understand its dynamics in the household sector, and thus pave way for improved and informed technology policies and programmes, as well as deployment strategies into the future [2].

This segment of the book focuses on the rationale supporting the introduction of prepaid electricity meters in Sub-Saharan Africa. Part 1 sets the context for upcoming discussions on prepaid electricity meters. It sets the scene by tracing the three decades historical experience of the continent with the technology; with South Africa being the first in 1988 and Angola being one of the youngest users of the technology. The objective is to trace the gradual expansion of the market as it replaces conventional or post-payment electricity meters. With that done, Chapters 3 and 4 extensively probes the effect of the technology within households with poor socio-economic conditions.

2.2 From Conventional to Prepaid

All electrified households are fitted within some payment model via a particular technological system. There are currently two universal payment models: Post-payment (credit) or smart prepaid model. These systems enable the consumer to quantify and pay for electricity consumed. A household using a post-payment meter model pays monthly bills according to what has been consumed. This means that the provision of electricity service precedes payment) [2, 4, 7–10, 14, 20–22, 25, 26]. This model of payment involves several administrative or operation activities, which include the

© The Author(s), under exclusive license to Springer Nature Switzerland AG 2021
N. Kambule and N. Nwulu, *The Deployment of Prepaid Electricity Meters in Sub-Saharan Africa*, Lecture Notes in Electrical Engineering 759,
https://doi.org/10.1007/978-3-030-71217-4_2

Table 2.1 Smart prepaid electricity meter challenges for the utility and consumer [2, 11, 12]

Utility	Consumer
Meter reading errors and therefore billing irregularities	Meter reading errors and therefore billing irregularities
Readings at times not accessible leading the meter reader to then estimate	Readings at times not accessible leading the meter reader to then estimate
Difficulty in managing usage because readings are not accessible	In case of non-payment utility may disconnect
Pilferage—illegal connections	Lumpy payment bills
Considerate amount of time between the meter reading, administration of and the delivery of accounts and the due date for payments	Processes of switching off the electricity due to non-payment can be problematic. (No warning is given beforehand)
Higher expenses because of the administrative tasks and logistical support for meter billing	Late delivery of post-paid bills to households
Late delivery of post-paid bills to households and therefore delayed revenue collection	Arbitrary electricity consumption
Constrained relations between supplier and client in the event of incorrect billion, during disconnection and reconnection	Constrained supplier and client relations in the event of incorrect billion, during disconnection and reconnection

delivery of electricity bills after meter readings and possible disconnection and reconnection of the consumer. Over the years, the system has become cost-ineffective for both the utility and the consumer. Consequently, several governments and utilities are reconsidering the widespread application of this payment model. Table 2.1, is an illustration of general issues associated with this system (may of course differ by geography and time).

2.3   Prepaid Electricity Meters: Technology Characterisation

Given these challenges, utilities and governments are resorting to the alternative payment model or technological innovation—a prepaid electricity model or technology (Fig. 2.1). Under a pre-payment model for electricity payment transaction precedes the consumption or provision of electricity service. So, households use only what they can afford to purchase. The consumer pays for electricity services in advance before consumption [2, 7, 8, 15, 21]. Households thus 'hold credit and then use the service until the credit is exhausted' [22, p. 3]. It is a 'technological tool that plays the role of mediator between energy-producing agents and consumers' [15, p. 241].

There are various types of prepaid electricity meter systems in the world: namely: keypad based systems, disposable card systems (one-way) and two-way smart card systems (Table 2.2).

Fig. 2.1 Prepaid electricity
meter tool

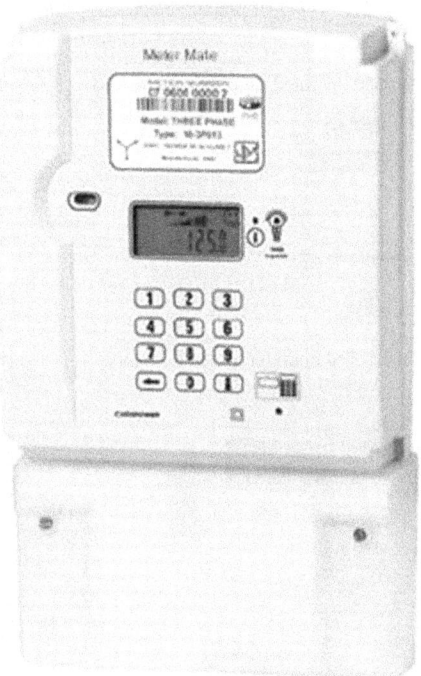

Normally, the prepaid electricity meter device is located within the household, thus enabling the consumer to easily access regular feedback on the quantity of electricity consumed [11, 12]. The device works as a mobile phone system. One has to top-up or recharge the system for it to fulfill its function—that is to activate the supply of electricity in the house. The consumer has to purchase electricity units and get a token with a reference code, which will be digitally entered into the device and converted into electricity units (kWh). When the numbers are entered correctly the system will automatically activate. The amount of credit loaded shows on the device screen (as kWh) and the end-user is therefore enabled to occasionally monitor electricity consumption and accordingly adjust behaviour [15].

The prepaid electricity meter is characterised by a unique regulatory option which primarily motivates consumers to pay for electricity services delivered [5, 20, 21]. Furthermore, the following technical and operation components have to be in place in order for the system to functioning efficiently [1, 27]:

- Prepaid electricity meter: Electricity Dispensers (ED) instrument
- Vending machines: to access Credit Dispensing Units (CDUs).
- Data Concentrators (DCs): used to manage and collect transaction data from CDUs, also called the System Master Station.

The consumer may utilise the electricity credits until they are exhausted, thereby be disconnected [8, 27]. The costing and payment systems vary by geography and

Table 2.2 Types of prepaid electricity meter system [11]

Type	Description
Keypad operated prepaid meters	Electronic prepaid energy meters came initially with keypad systems for inputting the credit. Security of keypad payment system is very low. The main reason is that the algorithm of key creation is stored inside the meter and is available to hackers. Keypad systems were created when highly secure smart card payments did not exist. Although keypad systems are getting obsolete, it may still be cost-effective for remote villages, where two-way vending may not be feasible
Smart card operated prepaid meters	Customers may buy a reusable power debit card for the amount of energy they desire. These special, easy to use cards are individualized, keyed to each customer's meter and account number. The customer simply passes the card a few inches in front of the meter, and using an integrated card reader, the meter is reset to the number of kilowatt-hours contained on the card. The card also captures data critical for load forecasting Smart card operated meters can be used either as prepaid or post-paid. In such cases, each consumer is assigned 50-day credit. After each month, the consumer has to recharge a card to pay off his negative balance. After the full payment, his/her 50-day credit is restored. Credit can be time-based or amount-based. A mixed option like 50-day but not more than certain kWh is also possible
Prepaid meter using two-way fixed network automated meter reading	This is a real-time two-way communication facility, where the transaction data is stored on a central computer and authorized in the same manner as for credit card terminals. Here security is not at risk as the credit information is stored on the central computer. However the cost of implementing the AMR (Automated Meter Reading) scheme needs to be justified on other merits than for prepayment advantages

application. In some areas, the system generally functions within 60 amps; this may differ according to household needs. For example in South Africa, the household electrification vacillates between 20 and 60 amps.

2.4 Historical Development and Market Expansion

South Africa and the United Kingdom (UK) are two pioneering prepaid electricity meters countries in the world. Figure 2.2 shows the historical timeline of the technology of Sub-Saharan Africa concerning other selected global countries. South Africa is the oldest prepaid metered country in the region—it stands the first developing country, 31 years ago, to use the technology. Ghana, Rwanda, and Tanzania are among a few regional countries that have more than two decades of experience with the system. Since the first installation in South Africa, there is widespread adoption of the technology. Utilities across the continent are widely adopting prepaid electricity meters, as opposed to conventional systems.

The Mozambique household sector is currently the biggest (80%), followed by South Africa with close to 70% prepaid electricity meter users, in Sub-Saharan Africa (Table 2.3). In that particular country, the technology has elicited a number of favourable conditions since the introduction of the technology, including improved electrification rising from 5% in 2001 to 18% in 2011; increased revenue levels, from 88% in 2001 to 97% in 2011; an overall decrease in losses, from 43% in 1995 to 21% in 2011 [7]. Positive results as these have fuelled governments in the region to intensify technology roll-out.

Table 2.3 in the following section, shows the motivating factors responsible for this. To date, the technology makes-up more than 30% of the regional household electricity metering market, and the share is estimated to surge to 53% within the next coming 5 years [19]. It is estimated that by the year 2034, Sub-Saharan Africa would have experienced a 234% growth in the prepaid electricity market [19]. Later discussions will provide an elaborate discussion on the status of the expanding market and also possible ramifications associated with this.

In the past years, several scholars have dedicated time to studying and recording country experiences with the technology in Sub-Saharan African households. Table 2.4 reflects examples of prepaid electricity meters studies by country in Sub-Saharan Africa between 2003 and 2019. From the 17 countries that were notably using the technology, there was difficulty in sourcing credible research for 4 countries (Mali, Cote d'Ivoire, Sudan, and the Democratic Republic of Congo).

2.5 Highlights

- South Africa is the first country in Sub-Saharan Africa to utilise prepaid electricity meters. Globally, together with the UK, it is recognised as a pioneer in the application of this type of technology.
- The Sub-Saharan Africa prepaid electricity meter market is among the fastest-growing across the globe. Estimates project the expansion rate will reach ≈234% by 2034.

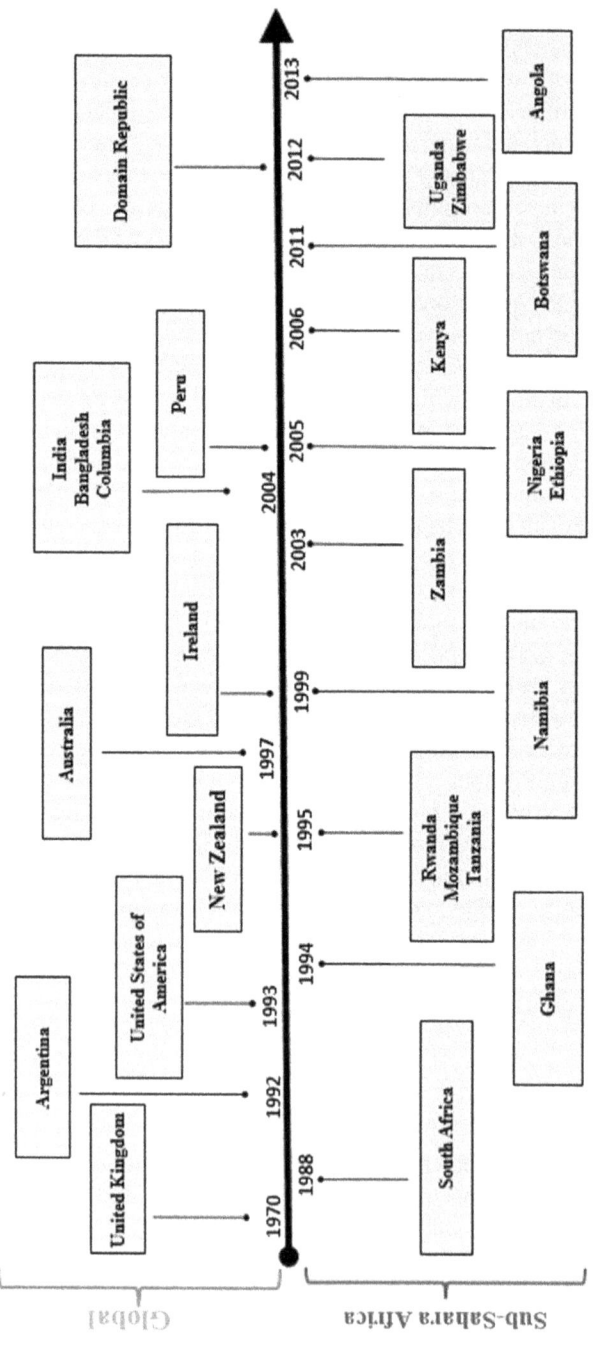

Fig. 2.2 Prepaid electricity meter historical timeline for selected countries of Sub-Saharan Africa (Authors)

Table 2.3 Prepaid electricity meter use statistics in Sub-Saharan Africa [7, 28]

Country	Quantified adoption status of adoption
Mozambique	80%
South Africa	66%
Tanzania	40%
Ghana	44%
Ethiopia	5%
Uganda	More than 14, 000 customers in rural areas
Rwanda	More than 90,000 customers a
Angola	Aims to install 6, 000 systems
Botswana	More than 298, 989 customers

Table 2.4 Prepaid electricity meter research studies in Sub-Saharan Africa (2003–2019)

	Countries	Researcher
1.	Uganda	Mwaura [18]
2.	Ghana	Azila-Gbettor et al. [3], Quayson-Dadzie [24]
3.	Mozambique	Baptista [4]
4.	Kenya	Miyogo et al. [17], Wambua et al. [30]
5.	Zambia	Malama et al. [15]
6.	Motswana	Mburu and Sathyamoorthi [16]
7.	Nigeria	Makanjuola et al. [13], Oseni [23], Aliu [2]
8.	Zimbabwe	Vutete [29]
9.	Tanzania	Jacome and Ray
10.	South Africa	Tewari and Shah [27], Kambule et al. [11, 12]
11.	Rwanda	Mwaura [18]
12.	Angola	Esteves et al. [7]
13.	Ethiopia	Esteves et al. [7]
14.	Mali	No research found
15.	Sudan	No research found
16.	Democratic Republic of Congo	No research found
17.	Cote d'Ivoire's	No research found

- Mozambique is the largest ($\geq$80%) prepaid electricity meter market in Sub-Saharan Africa, followed by South Africa ($\approx$70%).
- Across different regions the technology has managed to elicit several favourable conditions such as improved revenue collection for utilities and savings for households.
- There is still paucity with research focusing on the expansion of the prepaid electricity meter market and its role in the electrification programme.

References

1. ABS Energy Research (2007) Prepayment metering report. http://www.energy-market-res earch.info/publisher/ABS-Energy-Research.shtml. Accessed 16 Feb 2017
2. Aliu IR (2020) Energy efficiency in postpaid-prepaid metered homes: analysing effects of socio-economic, housing, and metering factors in Lagos, Nigeria. Energy efficiency. https://doi.org/10.1007/s12053-020-09850-y
3. Azila-Gbettor EM, Atatsi EY, Deynu F (2015) An exploratory study of effects of prepaid metering and energy related behaviour among Ghanaian household. Int J Sustain Energy Environ Res 4:8–21
4. Baptista I (2013) Everyday practices of prepaid electricity in Maputo, Mozambique. InSIS work paper series, pp 1–30. http://www.insis.ox.ac.uk/fileadmin/InSIS/Publications/Baptista_InSIS_WorkingPaper.pdf
5. Casarin AA, Nicollier L (2009) Prepaid meters in electricity. A cost-benefit analysis, working paper series IAE. https://www.iae.edu.ar/pi/Documentos%20Investigacin/Working%20Papers/DT%20IAE01_2009.pdf
6. Depuru SSR, Wang L, Devabhaktuni V (2011) Smart meters for power grid: challenges, issues, advantages and status. Renew Sustain Energy Rev 15:2736–2742
7. Esteves GRT, Oliveira FLC, Antunes CH, Souza RC (2016) An overview of electricity prepayment experiences and the Brazilian new regulatory framework. Renew Sustain Energy Rev 54:704–722
8. Franek L, Šťastný L, Fiedler P (2013) Prepaid energy in time of Smart Metering. Int Federat Autom Control. https://doi.org/10.3182/20130925-3-CZ-3023.00015
9. Jack BK, Smith G (2015) Pay as you go: pre-paid metering and electricity expenditure in South Africa. https://sites.tufts.edu/kjack/files/2015/08/Jack_manuscript7.pdf. Accessed 15 Nov 2016
10. Jack BK, Smith G (2016) Pay as you go: pre-paid metering and electricity expenditure in South Africa. https://sites.tufts.edu/kjack/files/2015/08/Jack_manuscript7.pdf. Accessed 15 Nov 2016
11. Kambule N, Yessoufou K, Nwulu N (2018) A review and identification of persistent and emerging prepaid electricity meter trends. Energy Sustain Dev 43:173–185
12. Kambule N, Yessoufou K, Nwulu N, Mbohwa C (2019) Exploring the driving factors of prepaid electricity meter rejection in the largest township of South Africa. Energy Policy 124:199–205
13. Makanjuola NT, Shoewu O, Akinyemi LA, Ajose Y (2015) Investigating the problems of prepaid metering systems in Nigeria. Pacif J Sci Technol 2:22–31
14. Makonese T, Kimemia DK, Annergarn HJ (2012) Assessment of free basic electricity and use of pre-paid meters in South Africa. http://conferences.ufs.ac.za/dl/Userfiles/Documents/00000/577_eng.pdf. Accessed 13 Oct 2016
15. Malama A, Mudenda P, Ng'ombe A, Makashini L, Abanda H (2014) The effects of the introduction of prepayment meters on the energy usage behaviour of different housing consumer groups in Kitwe, Zambia. AIMS Energy 3:237–259
16. Mburu PT, Sathyamoorthi CR (2014) Switching from post-paid to pre-paid models: customer perception and the organisational role in managing the change: a case study of botswana power corporation. J Manage Res 3:175–189
17. Miyogo NC, Ondieki NS, Nashappi GN (2013) An assessment of the effect of prepaid service transition in electricity bill payment on KP customers, a survey of Kenya Power, West Kenya Kisumu. Am Int J Contemp Res 9:88–97
18. Mwaura FM (2012) Adopting electricity prepayment billing system to reduce non-technical energy losses in Uganda: lesson from Rwanda. Utility policy 23:72–79
19. Northeast Group (2014) Sub-Saharan Africa electricity metering: market forecast (2014–2024). Northeast Group, LLC
20. O'Sullivan KC, Howden-Chapman PL, Fougere G (2011) Making the connection: the relationship between fuel poverty, electricity disconnection, and prepayment metering. Energy Policy 39:733–741

21. O'Sullivan KC, Viggers HE, Howden-Chapman PL (2014) The influence of electricity prepayment meter use on house hold energy behaviour. Sustain Cities Soc 13:182–191
22. O'Sullivan KC, Howden-Chapman PL, Fougere G (2015) Fuel poverty, policy, equity in New Zealand: the promise of prepayment metering. Energy Res Soc Sci 7:99–107
23. Oseni MO (2015) Assessing the consumers' willingness to adopt a prepayment metering System in Nigeria. Energy Policy 86:154–165
24. Quayson-Dadzie J (2012) Customer perception and acceptability on the use of prepaid metering system in Accra West region of the electricity company of Ghana. http://dspace.knust.edu.gh/bitstream/123456789/4900/1/Quayson-Dadzie%20John.pdf. Accessed 16 Oct 2017
25. Szabo A, Ujhelyi G (2014) Can information reduce non-payment for public utilities? Experimental evidence from South Africa. http://ftp.repec.org/opt/ReDIF/RePEc/hou/wpaper/2014-114-31.pdf
26. Taale F, Kyeremeh C (2016) Households' willingness to pay for reliable electricity services in Ghana. Renew Sustain Energy Rev 62:280–288
27. Tewari D, Shah T (2003) An assessment of South African prepaid electricity experiment, lessons learned, and their policy implications for developing countries. Energy Policy 31:911–927
28. Tuffour M, Sedegah DD, Asante K, Bonsu D (2018) The role of pre-paid meters in energy efficiency promotion: merits and demerits in Accra, Ghana. Int J Eng Trends Technol 1:57–64
29. Vutete C (2015) Adoption of the prepaid electricity meter billing system by harare residents: was there some preference to conventional meters? J Bus Manage 9:78–84
30. Wambua AM, Kihara P, Mwenemeru KH (2015) Adoption of prepaid electricity metering system and customer satisfaction in Nairobi County, Kenya. Int J Sci Res 9:1702–1710

Chapter 3
Rationale Part II: A Misdiagnosis of Non-payment and Electricity Theft

There is an underexplored relationship between the prepaid electricity meters and the effect that has on households that are non-payers (for electricity services). In this chapter we refer to the phenomenon of electricity non-payment as *arrearage*. Public utilities in sub-Saharan countries like Uganda, Botswana, Zambia Ghana, Angola, South Africa, Nigeria, Ethiopia, and Mozambique are facing the challenge of household electricity arrearage (i.e. non-payment of the electricity service). In all these regional countries, prepaid electricity meters have been deployed to directly reverse this mounting problem [9, 12, 53, 58, 60, 63, 69, 76, 85, 87]; Kambule et al. [44–46]. The technology thus largely stands as the cost recovery mechanism for utilities and/or governments.

While there remains a wide lacuna in the quantification of the effect of household prepaid meter deployment in the sub-Saharan region, the ideal and direct result of increased prepaid electricity meter deployments is a decrease in the overall electricity debt. So, generally utilities stand to benefit from the technology. But can the same conclusion be drawn about the effectiveness of the technology on households, particularly poor households based in the Sub-Saharan Africa region? This question becomes critical to consider especially in view of the fact that any contemporary form of development is framed and measured by its nature of sustainability.

Utilities have the legal obligation of disconnecting any household that does not pay for the electricity services, but this right has been limitedly applied. Instead, prepaid electricity meters seem to be imposed on non-paying, if not all, households. Tied to the issue of non-payment is that of electricity theft. It is estimated that by 2040, the global annual electricity consumption would have surged by 70% as compared to 2015 [42]. Faced with the reality of deepening poverty and inequality, it is important to consider the possibility of a surge in household electricity theft, especially in the context of the prepaid electricity meter. Is there a way that public utilities in the Sub-Saharan African region optimally deal with these challenges?

This chapter focuses on the phenomenon of household electricity arrearage and theft in Sub-Saharan Africa; particularly, in relation to the introduction of prepaid

© The Author(s), under exclusive license to Springer Nature Switzerland AG 2021
N. Kambule and N. Nwulu, *The Deployment of Prepaid Electricity Meters in Sub-Saharan Africa*, Lecture Notes in Electrical Engineering 759,
https://doi.org/10.1007/978-3-030-71217-4_3

electricity meters as a cost-recovery machine for utility and governments. We probe if the technology is the appropriate prescription. What is the real diagnosis of arrearage and electricity theft? Are prepaid electricity meters the solution to these problems? We argue that so far, our region has been shooting aimlessly, and there is therefore a need for a more focussed and holistic approach to dealing with these problems in a prepaid electricity metered marked.

3.1   Arrearage

While there are other issues that prepaid electricity meters solve (e.g. improved accessibility, reduced non-technical issues, etc.), the main motivation thus far has been to reduce electricity arrearage (Table 3.1). International studies and confirmed that the installation of prepaid electricity meter is for reducing household non-payment [9, 15, 17, 19, 22, 30, 31, 33, 35, 37, 49, 54, 55, 70–74, 77].

Electricity arrearage is a ubiquitous phenomenon plaguing the global electricity sector [2, 50]. Countries like the United States of America (USA), United Kingdom (UK), Eastern Europe, New Zealand, Australia, Argentina, Ireland, Colombia, Peru, India, South Africa, Botswana, Colombia, Kosovo and Soviet Union are examples of countries with households that have unpaid bills. Moreover, a correlation between non-payment and expenditure ratios (i.e. the greater the electricity consumption as

Table 3.1   Motivating factors for adopting prepaid electricity meters [44]

Country	Motivations for the introduction of prepaid meters
South Africa	• Increase the electrification rate • Reduce arrears • Reduce non-technical electricity losses and fraud
Mozambique	• Increase electrification rate • Reduce non-technical electricity losses and fraud • Solved efficiencies faced on meter reading and billing • Reduce bad debt level • Reduce the number of customers without meters
Nigeria	• Power sector re-organization • Solve billing and revenue problems • Reduce Distribution system operator (DSO) non-technical electricity losses
Ethiopia	• Solve billing and revenue problems • Improve the quality of service
Uganda	• Reorganization of the power sector • Reduce losses and non- payment
Angola	• Alleviate issues of household electricity debt
Ghana	• Address supply-side problems • Reduce operational costs • Improve revenue collection • Eliminate bad debt

a percentage of total household expenditure, the likelier the household does not pay its electricity bills) has been acknowledged [59].

Consumer payment for electrical services is pivotal for economic and infrastructural sustainability of electricity supplying utilities. The revenue is *inter alia* used for fixed development, operational and maintenance projects, research undertakings, and administrative services. The unavailability of funds due to arrearage may translate into a chain of intercompany arrears, bank arrears, tax and wage arrears—ultimately leading to maintenance backlogs, system deterioration, lack of funds to buy fuel to operate the generating units, and the deterioration of the economy [18]. Szabo and Ujheyli [84] write that non-payment for public-utilities is an important constraint that deprives expansion in the provision of electricity services.

The inability of households to pay for electricity services—engendering accruing household electricity debt—may be driven by different factors. In South Africa, the reasoning has a political element attached to it. In 1976, the African National Congress (ANC) of South Africa and other anti-apartheid political parties, fought against the ruling government by encouraging households to boycott the payment of electricity services. Moreover, as a means of political campaigns people were promised free electricity. Till today, people still believe that they are entitled to free electricity because of the historical promise made by the current ruling government—the ANC.

Free electricity is a non-real factor; household electricity arrearage is as such recognised as a criminal act [34]. Electricity arrearage (non-payment) is the inability—wittingly or unwittingly—of the user to pay for the electricity consumed. Szabo and Ujhelyi [84] write that,

> Improving people's access to basic utilities like electricity is viewed as a key challenge in many developing countries. However, consumers' ability or willingness to pay for services can be an important constrain to investment in infrastructure.

Furthermore, an economic perspective to non-payment may lead to one conclusion, a social perspective to the non-payment may warrant a different one. Electricity non-payment, from a social perspective, requires a consideration of aspects such as employment status, demographics, and literacy among others. Socio-economic conditions are arguably and normally the underlying cause of this scourge of non-payment [19, 45]. The anticipation would therefore follow that utilities in countries characterised by poor socio-economic realities would be faced with this problem more than those based in economically developed regions. While utilities have the power to cut-off electricity supply to non-paying electricity users, this option has not been popular in most African regions, instead the adoption of prepaid electricity meters has been the ultimate recourse.

The implication of this recourse is two-fold. The first implication should address the obvious question: *Is the utility recovering the cost, as intended through the technology—therefore is household arrearage mitigated?*

Research has shown that prepaid electricity meters do achieve their goal of reducing debt, thus is of benefit for the utility. For example, it was found that in

Zambia electricity arrearage dropped by close to 80% in one year through the adoption of prepaid electricity meters. Again, South Africa is estimated to be saving close to 2 billion per annum from the technology, revenue loss decreased by more than to 14% [86]. However, this positive aspect should be studied with caution. A township in South Africa, Soweto is a quintessential case to consider household non-payment. In 2015 municipalities across the country owed the utility $\approx$R13.5 billion due to non-payment of bills [83]. Soweto, alone, owed R8.6 billion of this amount. Estimates show that by 2019 the township's debt had escalated to about R10 billion. This is consistent with the countries risen overall household electricity debt of R26 billion in the same year. This is despite the expanding prepaid electricity meter market.

In the above case, the ratio of arrearage increase remains stronger than that of mitigation. This reality makes it clear that there is a misdiagnosis of the problem and/or a misprescribed solution. Utilities have blindly and ignorantly assumed that the problem is arrearage; this is a secondary problem that is a consequence and not the essence.

A stronger and often unobvious implication or question is overlooked and hence the misdiagnosis and misprescription—that is, *in the context of entrenched poor socio-economic conditions, what technology deployment strategy can be employed to elicit beneficial livelihood conditions.* Scholars such as Colton [19], O'Sullivan et al. [71, 73, 72], and Kambule et al. [45] have ascertained the negative socioeconomic implications associated with prepaid electricity meters. Chapter 4 expounds further on this point. However, it is apparent that there is a misalignment in the nature of the deployment prepaid electricity meters approach and the widespread household socio-economic livelihood.

3.2 Disconnection: An Ignored Alternative?

In the foregoing section it was established that a natural response to non-payment— disconnection or denied service. While this should be the case but until now utilities and governments have applied this natural alternative limitedly. This is due to the fact that it is often not feasible in Sub-Saharan Africa and other developing countries to just cut-off non-paying households [84]. A natural response may be in contrast to social ideologies of welfare and fairness and therefore erode citizen's trust in government; ultimately resulting in social unrest and more non-payment. When the consumer refuses to pay because of socio-economic reasons, the utility may find it difficult to take any enforcement action against the consumer. Utilities therefore have limited enforcement power, if any, it is often costly. To prevent this, countries need to deal with the issue with necessary sensitivity and precaution.

Between 1994 and 2001, approximately 10 million South African households experienced electricity and water disconnections due to non-payment. About 95% of electricity connections in the country are for households–making them important electricity customers (even from a cost-recovery point of view). Despite having one of the lowest residential tariffs in Sub-Saharan Africa, Eskom has voiced concerns about the prevailing electricity payment inefficiency from the municipalities and/or end-user, resulting in accruing utility debt. In 2015, households owed the Eskom (the electricity supplying utility) more than R13.5 billion. Earlier that decade, around 2003, the household electricity debt of Soweto was around R1.4 billion due to arrearage. The utility erased that debt under the Integrated National Electrification Programme (INEP) to encourage customers to pay for the service. Irrespective, by 2015 the amount owed had risen again, this time to ±R10.6 billion [83, 86]. Until today, less than 16% of households in Soweto pay for electricity services. Eskom has a legal obligation to directly disconnect the 84% of its non-paying customers it has thus far applied this approach limitedly and has however resorted to prepaid electricity meter installation as a means of mitigating debt and household electricity arrearage.

Another case of non-payment is that of Georgia in America. In 1998, only about 15–38% of the electricity generation capacity was operational, and as such households received electricity for about 6 h per day [50]. The lack of investment hampered efforts for infrastructural development and maintenance. The lack of capital was due to non-payments, over-subsidization, and electricity theft contributed to the low cost-recovery. Higher tariffs (affordability), free-riding, political tolerance of non-payment, lack of incentives on the part of corporate management to resist political pressure, lack of high-level political commitment, weak enforcement of laws and regulations, theft, corruption, and falling household incomes also contributed to non-payment in the region [81].

Borrowing from an international example, we consider Kosovo. In this country electricity non-payment accounts for about 44% of electricity consumed by households—this translates to nearly €100 million per annum [84]. The main reason for non-payment relates to household economic problems. Household electricity bills consume more than 30% (even higher in winter) of the total share of monthly disposable income. The residents also argue that the Kosovo Energy Distribution Services (KEDS) is not transparent and charging more than it should. Group thinking is another contributing to non-payment in area, wherein citizens find non-payment reasonable because other households are doing the same. More than 81% of households (Albanians) in Kosovo indicated that despite not having paid for their electricity bills, they were yet to be disconnected by the utility.

From what has been noted above, causal factors of household non-payment can be summated as outlined below:

- Declining incomes (salary or wages), rising unemployment, and escalating electricity tariffs.
- The inability of the utility to enforce household disconnection and lack of political will to intervene.
- The inability of the utilities to disconnect supplies to non-paying customers, as governments maintained long lists of strategic consumers to which supplies could not be disconnected and zealous local politicians went on adding to this list several local industries to protect local jobs, and the local economy at the expense of the energy firms.

- In most countries, energy expenditures account for 15–30% of per capita income, as such a majority of households find it difficult to pay bills.
- Limited electricity infrastructural development.

Notably, countries experience the problem of non-payment in different ways and therefore have different solutions to the problem. Table 3.2 provides an outline of deployed mechanisms to alleviate household non-payment. Regions either use one or a combination of mechanisms.

From the cases considered above, non-payment is both an economic and social issue. From a household perspective, whereas non-payment may be a free-riding for some, for others electricity consumption is largely an aspect beyond the means of control because of socio-economic factors at play. Due to this dynamism, to understand and unearth the real cause, and therefore devise a potentially sustainable solution, an objective inquiry is important. Lampietti et al. [50] recommend that "the interests of the government and the utility need to be aligned to back reform and share the risk of non-payment". In a region like Sub-Saharan Africa, with deep inequality realities, a blanket approach to solving the issue may benefit some and still compromise the other [3].

Household electricity non-payment is a systematic problem and is therefore multi-dimensional, characterised and driven by a variety of factors (i.e. economic or financial, social, infrastructural, and regulatory). This means that in devising a sustainable solution to curb non-payment there is a need to equally consider different factors. The diagnosis may be different by country, and therefore also the prescription will be different. Thus far, the prepaid electricity meter programme or solution in the Sub-Saharan region has primarily been cost recovery or economic oriented. While there may not be a problem with the technology per se, but how the technology is deployed across the region should be preceded by thorough socio-economic assessments. Failure to undertake this will inadvertently breed an environment for energy poverty (Chap. 4).

3.3 Power System Losses

Energy loss through the grid network is an inevitable phenomenon. Countries across the world, both developed and developing regions experience this problem, though to different extents. The losses may happen along the distribution or transmission networks [1]. The distribution losses are used as an important performance indicator as they have a direct bearing on the economy of the utility. Distribution losses are "the difference between the amount of energy delivered to the distribution system and the amount of energy biilled to the customers" [16]. Ideally, the amount of energy generated should be equal to the energy registered to have been used by the end-user. Figure 3.1 gives an illustration of how these system power losses are incurred.

These losses can further be divided into two, namely technical and non-technical losses. While technical losses are recognised as inherent (i.e. from Dielectric losses,

Table 3.2 Generic mechanisms to deal with non-payment [41, 40]

Mechanism	Description	Example of region
Communal metering	The community bears a collective task for paying the electricity consumed and therefore the bill. The method is more effective than individual household meters when there is a threat of a whole community electricity disconnection a whole intention is increasing collections and keeping tariff costs low. Community self-policing is also key Communities are metered at a single remote location (where tampering is difficult) and all customers are billed based on a consumption determined by pro-rating using their meter reading. This acts to deter offenders through direct community pressure	USA, UK, Georgia, Russia, Uganda
Bill fraction-payment	This is more of a coping mechanism where the non-payer resorts to only paying a fraction of the bill and maintaining service while accumulating arrears	South Africa, Armenia, USA and other countries
Monitor electricity consumption	Consumers closely monitor the amount of electricity used and then self imposed austerity measures when the limit is reached	All countries encourage this
Prepaid meter system (Split metering)	The installation of a prepayment system allows the consumer to pay for electricity in advance as opposed to a credit meter system where electricity is consumed before a payment transaction. A split meter system uses two electricity recording devices, one placed in-house and another outside the house (mounted on pole or pavement); this ensures that there is no interference between the utility and the consumer	South Africa, UK, USA, Australia, China, Brazil, New Zealand, etc.

(continued)

Table 3.2 (continued)

Mechanism	Description	Example of region
Enforcement	Relevant institutions, like the government, may need to improve law enforcement for payment of electricity. Enforcement improved collections—Russia, Georgia, Ukraine, and Armenia. The legal infrastructure, may need to properly the rights of the energy supplier (e.g. to deny supplies to those who do not pay for it), and the consumer (e.g. protection from illegal disconnections)	All countries (to) encourage this
Energy transfer programs	A certain ('deserving') population category (e.g. veterans, pensioners, poor, etc.) is allocated with electricity at defined intervals	South Africa, Armenia, Britain Ireland, UK,
Elimination of unmetered consumption	This is a utility intervention whereby meter relocation to places accessible to utility staff is necessary. The benefit of the intervention is theft prevention and improving payment collections—Armenia, Albania, and Georgia	South Africa
Electricity Price discounts	This is an incentive to encourage consumers to pay electricity	UK
Education	Information is noted as important to understanding how the electricity service sector functions. From this flows that the lack of familiarity with the consumption billing process is the cause of non-payment. Education may happen through public information campaigns which may generate unintended consequences such as psychological responses and improve the payments	South Africa, Uganda

Copper losses, and Induction/radiation losses) and be avoided by improving the extant infrastructure and technology, the extent of non-technical losses often exceeds the technical losses experienced [16]. A non-technical loss refers to commercial losses that comprise of energy that has been "delivered and consumed energy which cannot be invoiced to an end-user" [25]. Bula et al. [16] add that the losses are due to

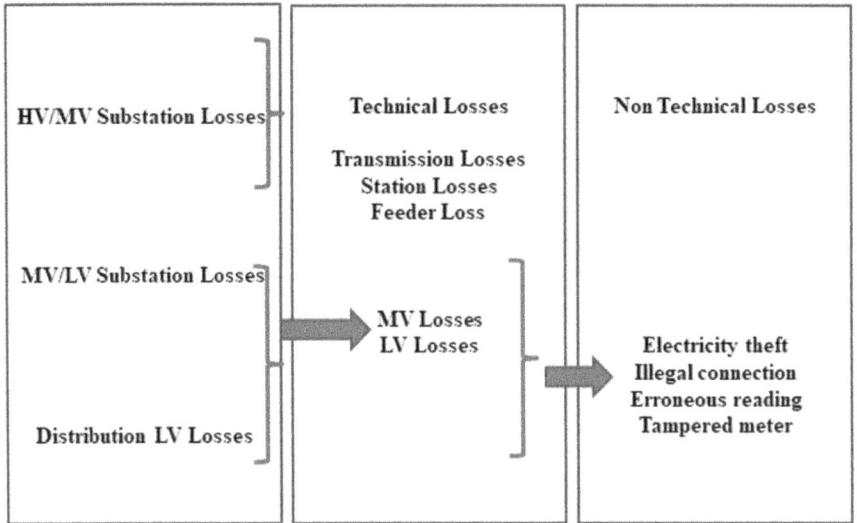

Fig. 3.1 Types of system power losses [7, 89]

unauthorised means of consuming energy via unauthorised connections to the meter. In an ideal setting, the customer receives a bill from the utility on the amount of electricity used and to be paid for. However, under non-technical losses most of the energy wrongly received by the consumer is unaccounted for. This ultimately affects the economy of utilities. In the discussions below we will give examples of the extent of losses suffered by utilities. Generally, the losses affect the operating costs of the utility, ultimately leading to an increase of consumer electricity tariffs, and may also lead to an increase in government subsidies [57, 89]. The problem ultimately affects investments for capacity development and eventually leading to electricity shortage. This is particularly a problem in Sub-Saharan Africa, where power utilities experience marked inefficiencies [29].

New-technical losses can be divided into several types, including fraudulent losses that generally entail illegal connections, theft, meter tampering, and faulty meters. In other cases the losses may be due to the use of non-compliant material in technologies such as transformers. According to Mbanjwa [57] there are also hidden losses from in-house consumption of the equipment along the distribution network, this may be in cases where the power is needed to cool off a piece of technology such as a transformer and operate control systems. Among these types of non-technical losses electricity theft is by far the rifest and devastating [89]. This is a reality that again faces both developed and developing regions. Electricity theft is when an individual does not pay, or pays less of what they should, for the used electricity because they have bypassed or tampered the electricity meter [16]. The ability to reduce non-technical losses can translate into direct economic advantages for the consumer with reduced electricity costs and also for the utility with improved income inflow. Navani [65, p. 1] summarises the causes of non-technical losses in the following manner:

The most probable causes of Non-Technical Losses (NTL) are: (i) Tampering with meters to ensure the meter recorded a lower consumption reading (ii) Errors in technical losses computation (iii) Tapping (hooking) on LT lines (iv) Arranging false readings by bribing meter readers (v) Stealing by bypassing the meter or otherwise making illegal connections (vi) By just ignoring unpaid bills (vii) Faulty energy meters or un-metered supply (viii) Errors and delay in meter reading and billing (ix) and lastly, Non-payment by customers

However, we must look into how developing or emerging economies have dealt with or struggling to deal with this challenge of electricity theft. The next section discusses the extent, motivation, and consequences of electricity theft in the energy or electricity fraternity.

3.4 Electricity Theft

This is a specific challenge that haunts power utilities across the globe. It can be defined as an intentional attempt of the consumer to eliminate or reduce the cost reflected on the electricity as owed to the utility [13]. This would be categorised as illegal electricity consumption. Additionally, electricity theft is the "practice of using electricity from the utility company without the company's authorisation or consent [1, p. 1]. Technically, it is identified as the difference between the amount of electricity purchased from the power utility and that which is sold to consumers [57]. There are several ways that this can be done. It can be by directly connecting an unauthorised load to the network (or line) or tampering an already registered meter to reduce the amount of the bill the utility charges for that load [13, 16, 20]. Once the meter technology is vandalised, the system can be manipulated in different ways to either stop or slow it down. Table 3.3 shows the different ways that electricity theft can be undertaken [57].

Estimates show that electricity distribution companies globally annually lose about $25 billion on electricity theft [16, 48]. The North East Group [67] raises this estimate to be around $96 billion. To show how rife this issue is across the globe, consider the outline of the statistics below:

- India loses $4.5 billion annually on electricity theft. If they could reduce theft by 10%, they can annually save about 83, 000 GWh of electricity [16].
- The United States of America (USA) loses around $6 billion due to non-technical losses [16, 67]. Mbanjwa [57] asserts that theft costs the region between 0.5 and 3.5% of annual gross revenues approximately $280 billion (an annual loss of about $1.4 billion $9.8 billion per annum).
- One of Canada's electricity company (BC Hydro) spends $100 million annually because of electricity theft [1].
- Electricity theft rates in countries like Bangladesh and Turkey reach levels as high as 30% of the produced total electric energy generated [68].
- The cost of electricity theft in Brazil amounts to 4 billion annually [88].

Table 3.3 Types of electricity theft [5]

Type	Description
Electricity fraud	The consumer illegally uses electricity equipment and services with an intention of not paying and avoiding the electricity bill [21]. The consumer deliberately deceives the power supplying utility, this is commonly done through tampering with the meter so that lower than expected consumption levels reflect
Billing irregularities	Billing irregularities transpire in different ways. Inaccurate readings may be captured by the servicemen through receiving a bribe from the consumers. The bill is intentionally fixed in the offices in exchange for illicit money from households [24]. Again, power authorities may sometimes be unable to enter consumer premises to capture the amount of electricity utilised. Consequently, the serviceman gives an estimate which is much higher or much lower
Illegal connections	This is a connection that is undertaken illegitimately, that is without the knowledge, authorisation, and permission of the electricity supplier for that particular area (e.g. Eskom, ZESCO, etc.). Illegal connections involve the rigging of one network from a certain power source to where it will be connected (thus bypassing meter) [82]
Meter tampering	This is another form of theft whereby the electricity meter is bypassed in order to preclude the payment of electricity consumed
Unpaid bills	Unpaid bills can be defined as the omission or failure to pay for a particular service that is rendered by the utility. For example, in this case, the non-payment of electricity would be a result of households or organisations avoiding the payment of electricity bills [61]—ultimately breeding a culture of non-payment

- Malaysia and Lebanon experience lose 40% of electricity from non-technical losses, rendering peaking shortages endemic [21, 39, 78].
- South Africa suffers annual losses of about $1.5 billion on electricity theft and illegal connections. The non-technical losses show themselves as vandalism infrastructure, the abstraction of substation oil, cable theft, non-payment electricity, and the trading of illicit prepaid meter vouchers) [75]. Mbanjwa [57] adds that by 2004 Eskom had already been losing R250 million on electricity theft. By 2007, the losses amounted to a total of 6 105 GWh.
- In 2019, the city of Cape Town reported that 3 000 electricity cable theft and 6, 000 cases on electricity tampering [75].
- The cost of electricity theft at eThekwini municipality was more than R150 million in 2016 [57].
- Nigeria's electricity sector suffers annual non-losses of about $54 million annually (ESI [27, 32].
- Zimbabwe also annually loses about $237 billion (290 GWh) [62].
- Rwanda suffers losses of $200 million every year from theft. This is close to 10% of the total electricity generated in the country [79].
- Mozambique annually loses $100 million from non-technical-loses [36].

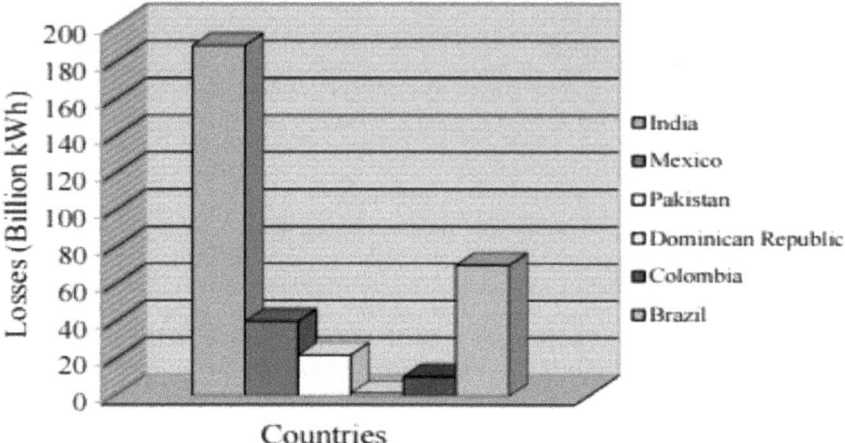

Fig. 3.2 Electricity theft and loss in kWh among selected developing countries [16]

Figure 3.2 reflects examples of unit (in kWh) loss that developing countries experience annually from electricity theft. India and Brazil are among the leading countries with more than 180 and 90 billion units stolen annually, respectively.

To apply the correct solutions it is important that we also know the different and specific motivations for this problem. As noticed, both developed and developing regions suffer from this problem. While developed regions between 3.5 and 30% of revenue to electricity theft, developing countries lose between 30 and 47% electricity is from theft [8, 11, 57, 62]. However, the motivations for theft are reportedly different [6, 39, 47]. In developing regions the prices of electricity are relatively higher and electricity theft therefore comes as a means of avoiding payment. But in developing regions where the tariffs are lower, general poverty drives consumers to steal [48, 57]. The report specifically notes that poor service delivery is an important contributor to theft. Customers are generally motivated by varieties of factors that are premised by the following [16]:

- Political or governance
- Socio-economic
- Managerial
- Educational
- Legal
- Managerial
- Infrastructural.

Of all the listed factors, the most prominent driver to electricity theft is socioeconomic. This factor influences the ultimate ability of the nature of individual access to electricity consumption. Specifically depicted in the figure below is a list of socioeconomic parameters or variables that have to be studied to understand the nature of electricity theft [10, 38] (Fig. 3.3):

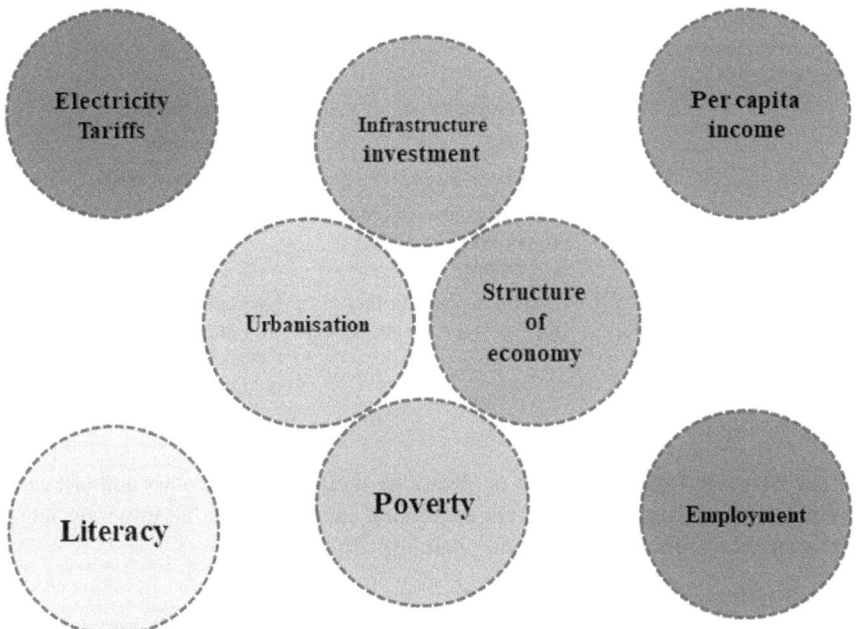

Fig. 3.3 Socio-economic variables for assessing electricity theft [10]

In the context of households, affordability may be an important factor in driving theft [23, 89]. However, this is not always the case. In a country like South Africa where the prices of electricity have in the past two decades been relatively low, electricity theft has equally been rife, so in cases like there may be a need for a more rigorous assessment on what drives the problem. Mbanjwa [57] advises that in dealing with electricity theft "it is imperative to examine factors other than income."

Other studies have also noted that that theft, for example in areas of India and Pakistan, is closely related to governance, wherein countries without effective accountability, political instability, and corruption have a higher rate of electricity theft [82]. In countries like Jamaica it has been found that theft is closely associated with low levels of education and income. Factors motivating for electricity theft rural regions are largely attached to the reality of underdevelopment, while in urban space this may be due to, though not exclusively, overpopulation. The table below reflects some of the fundamental motivations between rural and urban areas [10, 16, 47, 48, 56, 57] (Table 3.4). Reasons for electricity theft are generally the same for both rural and urban areas. In rural areas the difference is the poor implementation of the pro-poor policies. While in countries like South Africa, Uganda, and Rwanda have such policies in place but these mechanisms are not yielding the desired outcomes for poor households.

Table 3.4 Motivations for electricity theft—Rural and Urban areas [14]

Rural	Urban areas
Poverty	Poverty
Unemployment	Unemployment
Inequality	Social inequality
Inability to afford	Inability to pay bills
Poor implementation of the pro-poor policies or programmes	Poverty
Corruption public utility employees (i.e. taking bribes)	Easy access to an individual who can illegally bypass a meter
	Fraud and unpaid bills

The heavy and negative effects of electricity theft are what prompt authorities to find solutions to the problem. Before addressing and discussing the solutions, let us note some of the effects of electricity theft [10, 29].

- Overloading of the electrical system
- The inability of utility to predict accurate peak demand (which is important for decision making)
- Financial losses
- Increased electricity tariffs for the consumer
- Increased prices of essential commodities such as food
- No adoption of energy efficiency measures, leading to the abuse of electricity
- Threatens community safety and increases the risk of loss of life.

3.5 Mechanisms to Curb Electricity Theft—Is a SPEM the Solution?

The cost of electricity theft is too high for authorities to overlook the problem. As mentioned above, this criminal activity affects even investment opportunities. So authorities in the Sub-Saharan region have started to seek ways to alleviate this challenge. Below is a table that illustrates different solutions that authorities have adopted or can adopt in the attempt of curbing electricity theft in Sub-Saharan Africa (Table 3.5). The nature of the adoption of these should differ by country and context.

One of the motivations for increasing the use of prepaid electricity meters has also been due to the need to reduce the problem of electricity theft and improve utility revenue, hence the increasing deployment of smart SPEM in the region [62]. Between 2013 and 2017, through prepaid electricity meters Eskom managed to save about R1.4 billion ($109,000) in revenue by decreasing electricity theft by 0.69% (from 7.1 to 6.4%) [29, 75]. However, this decline, although important, but is insignificant. For instance, in South Africa where prepaid electricity meters were introduced in

Table 3.5 Mechanisms for dealing with electricity theft in the Sub-Saharan region [4, 8, 26, 82]

Solution	Description
Technical	In the recent past, the ability to detect electricity theft has emerged as an area of importance for research and development. This is as authorities and public utilities seek to develop mechanism of reducing electricity theft to find ways of dealing with this very economically costly and deadly practice. Government or utilities increasingly applying special devices such as smart technologies to detect and resolve theft issues [64]. These smart meters have been built with engineering algorithms that show electricity consumption patterns [80]. The software can assist in detecting electricity that is being consumed illegally. The challenge that comes with this method/solution especially in the developing Sub-Saharan region is the poor availability of skills to manage and monitor the system. Additionally, the technology remains expensive for most countries to easily access and deploy [82]
	Some countries, like South Africa, have started to use tamper-proof meters. These meters are made of iron that houses the actual meter inside. The iron is not easy to grind through They need a special key that only Eskom employees possess. In most township areas of South Africa, these boxes currently lay open and are only closed or sealed by cello tape. Households easily gain access to the iron box and system and therefore can manipulate the system to permit them to consume more and pay less
	Vigilant Energy Metering System (VEMS) with Advanced Meter Infrastructure (AMI) system is also another option that can detect anomalies in household electricity consumption patterns
	In some areas the use of Mobile Remote Check Meters is becoming popular. This method can however be used only to detect electricity theft at a micro scale (i.e. meters with low voltage) [62]
Awareness campaigns	Countries can establish civil organizations that monitor electricity theft in communities. *Operation Khanyisa* and *Vuk'uzenzele* are prominent organisations in the region that help utilities such as Eskom to track and reduce electricity in local communities
	More than that, the availability of media space can also be used as a means of raising awareness against electricity theft. In South Africa you find a special hotline that communities can use to report *izininyoka* (people that steal electricity). Governments and public utilities should take advantage of the television, radio, emails, and local newspapers to advertise and educate the general populace about the danger of electricity theft
	Education, through door-to-door engagements with community members or households can help utilities can a trusting relationship with households who can in turn help in reporting electricity theft. Some utilities in the region, as in South Africa, Tanzania, Rwanda, and Botswana already encourage its consumers to use available designed anonymous and affordable messaging services, and free hotline services [28, 66]

(continued)

Table 3.5 (continued)

Solution	Description
Governance and Managerial	Any solution can only be effective if and when there is political-will and transparent governance. In the case of electricity theft reduction, it can only be possible whereby the governance culture endorses law enforcement against this criminal act of electricity theft
	With the reality that in some instances internal management is centrally involved in facilitating electricity theft in communities, it therefore becomes pivotal that there be independent periodical managerial inspections and monitoring
Regulation, Enforcement, and Penalties	Strict regulations are necessary to reduce theft. More than that, the enforcement of present regulations is imperative. Mbanjwa [57] found that more than 50% of electricity theft cases are yet to be presented on a court roll [57]. In a different study, Lowvelder [52] ascertained that 96% of South Africans are aware that electricity theft is a criminal act. However, a mere 16% believe that they will get caught. This is an important indicator that there is a need for law enforcement
	Beyond a design of laws that will force authorities to enforce relevant pro-poor laws, for example South Africa's FBE policy. Currently no law exists in the Sub-Saharan region holding local authorities accountable for pro-poor policies
	Governments need to issue out financial penalties for any household found to have tampered with the meter (see Addition Table below). In some cases, houses opt to give bribes to management and avoid high penalties

Additional Table [75]

Domestic tamper fees

First offence (excluding the meter and other-cost)	R4, 561
New meter	R1, 886
Other (disconnection and reconnection)	R612
Second offence (excluding the meter and other-cost)	R7,059
Third offence (excluding the meter and other-cost)	R12, 159

2007 and the technology has constantly been bolstered even towards replacing the traditional prepaid meters with smart technologies, this level of reduction is minute. This is in line with Smith [82]:

> The disturbing evidence is that losses (and theft) appear to be increasing in an era of readily available technological means (metering, for instance) to lower non-technical losses.

So, we ask the question again: *Can prepaid electricity meter assist in curbing theft.* If the technology fails to improve the socio-economic livelihood of households, particularly those in low-income households, then electricity theft will not end. The general primary driver of theft in Sub-Saharan Africa is socio-economic. Therefore, prepaid meters can only assuage electricity theft when the prepaid meter programme fundamentally considers this point. Table 4.4 in the next chapter elaborates on what a contextual and effective prepaid electricity meter deployment framework should consider. Until the application of such a framework, households will constantly devise mechanisms of vandalising or bypassing installed prepaid electricity meters.

3.6 Highlights

- Governments across Sub-Saharan Africa have introduced the technology in order to deal, either by curbing or averting, the challenge of household electricity non-payment. Whereas disconnection to electricity supply should be a natural response to arrearage, governments and utilities have exercised this legal approach very limitedly.
- Prepaid electricity meters have the potential of reducing arrearage by up to 80% in one year.
- Electricity arrearage is both an economic and social issue. A holistic approach to the issue is necessary. Thus far, the perspectives have either been economic or social. Effective dealing with non-payment, through prepaid electricity meters, require a holistic understanding and approach.
- Albeit South Africa is saving close to 2 billion per annum from prepaid electricity meters—and the revenue loss has declined by $\geq 14\%$, households continue to owe the utility about R26 billion. There is a need for an improved programme—a holistic approach.
- Before deployment of prepaid electricity meters to deal with arrearage, authorities need to understand the context and causal factors.

References

1. Abdullateef AI, Slami MJE, Musse MA, Aibinu AM, Onusanya MA (2012) Electricity theft prediction on low voltage distribution system using auto-regressive technique. Int J Res Eng Technol 5:2277–4378
2. Ahluwalia MS (2002) Economic reforms in India since 1991: has gradualism worked? J Econ Perspect 16:67–88
3. Aliu IR (2020) Energy efficiency in postpaid-prepaid metered homes: analyzing effects of socio-economic, housing, and metering factors in Lagos. Energy Efficiency, Nigeria. https://doi.org/10.1007/s12053-020-09850-y
4. Amarnath R, Kalaivani N, Priyanka V (2013) Prevention of power blackout and power theft using IED. IEEE global humanitarian technology conference

5. Amin S, Schwartz GA, Cardenas AA, Sastry SS (2015) Game-theoretic models of electricity theft detection in smart utility networks: providing new capabilities with advanced metering infrastructure. IEEE control systems, 1. 10.1109/MCS.2014.2364711
6. Anderson R, Fuloria S (2011) On the security economics of electricity metering
7. Andrew S, Prachi S (2013) The political economy of electricity distribution in developing countries: a review of literature. Political Governance, UKAID
8. Antmann P (2009) Reducing technical and non-technical losses in the power sector. World Bank, Technical Report
9. Azila-Gbettor EM, Atatsi EY, Deynu F (2015) An exploratory study of effects of prepaid metering and energy related behaviour among Ghanaian household. Int J Sustain Energy Environ Res 4:8–21
10. Baker L, Philips J (2019) Tensions in the transition: the politics of electricity distribution in South Africa. Environ Plan Politics Space 37:177–196
11. Balasubramanya C (2014) Electricity theft: is smart meter a solution? http://www.slideshare.net/balasubramanyachandrashekariah/electricity-theft-is. Accessed 12 May 2020
12. Baptista I (2013) Everyday practices of prepaid electricity in Maputo, Mozambique. InSIS work paper series, pp 1–30.http://www.insis.ox.ac.uk/fileadmin/InSIS/Publications/Baptista_InSIS_WorkingPaper.pdf
13. Bella GD, Grigoli F (2016) power it up: strengthening the electricity sector to improve efficiency and support economic activity. International Monetary Fund: Working papers. April 2016. WP/16/85
14. Brueckner JK, Lall SM (2014) Cities in developing countries: fueled by rural-urban migration, lacking in tenure security, and short of affordable housing. Handbook of regional and urban economics
15. Brutscher PB (2011) Payment matters?—an exploratory study into pre-payment electricity metering. http://www.econ.cam.ac.uk/dae/repec/cam/pdf/cwpe1124.pdf
16. Bula I, Hoxha V, Shala M, Hajrizi E (2016) Minimising non-technical losses with point-to-point measurement of voltage drop between "SMART" meters. IFAC-PapersOnLine 49–29:206–211
17. Casarin AA, Nicollier L (2009) Prepaid meters in electricity. a cost-benefit analysis. Working paper series IAE. https://www.iae.edu.ar/pi/Documentos%20Investigacin/Working%20Papers/DT%20IAE01_2009.pdf
18. Chehore T (2014) South Africa: electricity pricing paralysing the poor. http://www.ee.co.za/article/south-africa-electricity-pricing-paralysing-poor.html
19. Colton DR (2001) Prepayment utility meters, affordable home energy, and the low-income utility consumer. J Afford House Dev Law 3:285–305
20. Czechowski AMK (2016) The most frequent energy theft techniques. IEEE
21. Dangar B, Joshi SK (2015) Electricity theft detection techniques for metered power consumer in Guvnl, Gujarat India. In: Paper presented at Clemson power system conference (PSC 2015)
22. Darby S (2006) The effectiveness of feedback on energy consumption: a review for Defra of the literature on metering, billing and direct displays, pp 1–21
23. Depuru SSSR (2012) Modelling, detection, and prevention of electricity theft for enhanced performance and security of power grid. Doctor of Philosophy Degree in Engineering Dissertation. The University of Toledo, USA
24. Dike DO, Obiora AU, Nwokorie EC, Dike BC (2015) Minimizing household electricity theft in Nigeria using GSM based prepaid meter. Am J Eng Res 1:59–69
25. EE Publishers (2008) Electricity theft and non-payment—Impact on the SA generation capacity crisis http://www.ee.co.za/article/electricity-theft-and-non-payment-impact-on-the-sa-generation-capacity-crisis.html. Accessed 22 Apr 2020
26. ESI Africa (2016) Kenya power intensifies surveillance on electricity theft. https://www.esi-africa.com/news/kenya-power-intensifies-surveillance-on-electricity-theft/. Accessed 4 May 2020
27. ESI Africa (2019) The big question: Is Nigeria embracing the 4IR? https://www.esi-africa.com/industry-sectors/future-energy/the-big-question-is-nigeria-embracing-the-4ir/. Accessed 15 May 2020

28. Eskom (2016a) Free basic electricity. Available http://www.eskom.co.za/news/Pages/Apr18. aspx. Accessed 11 May 2020
29. Eskom (2016b) Illegal connections can be deadly. http://www.eskom.co.za/news/Pages/Oct13. aspx. Accessed 12 May 2020
30. Esteves GRT, Oliveira FLC, Antunes CH, Souza RC (2016) An overview of electricity prepayment experiences and the Brazilian new regulatory framework. Renew Sustain Energy Rev 54:704–722
31. Faruqui A, Sergici S, Sharif A (2010) The impact of informational feedback on energy consumption—a survey of the experimental evidence. Energy 35:1598–1608
32. Fidelis C, Amadi J, Helen A (2019) Abating electrical power theft in Nigeria using smart meters and data analysis. Curtailing Energy Theft, 1–25
33. Fischer C (2008) Feedback on household electricity consumption: a tool for saving energy? Energ Effi 1:79–104
34. Fjeldstad OH (2004) What's trust got to do with it? Non-payment of service charges in local authorities in South Africa. J Modern Afr Stud 42:539–562
35. Franek L, Šťastný L, Fiedler P (2013) Prepaid energy in time of smart metering. Int Feder Autom Control. https://doi.org/10.3182/20130925-3-CZ-3023.00015
36. Frey A (2019) Mozambique power utility loses $100 M annually. https://clubofmozambique. com/news/mozambique-power-utility-loses-100mannually/. Accessed 11 May 2020
37. Gans W, Alberini A, Longo A (2013) Smart meter devices and the effect of feedback on residential electricity consumption: evidence from a natural experiment in Northern Ireland. Energy Econ 36:729–743
38. Gaur V, Gupta E (2016) The determinants of electricity theft: an empirical analysis of Indian states. Energy Policy 93:127–136
39. Glaucner P, Boechat AP, Dolberg L, State R, Bettinger F, Rangoni Y, Duarte D (2016) Large scale detection of non-technical losses in imbalanced data sets. In: Paper presented the seventh IEEE conference on innovative smart grid technologies (ISGT 2016)
40. Graham S (1997) Liberalised utilities, new technologies, and urban social polarisation: the UK experience. Eur Urban Region Studies 4:135–150
41. Graham S, Marvin S (1994) Splintering networks: cities and technical networks in 1990s Britain. https://www.ncl.ac.uk/media/wwwnclacuk/globalurbanresearchunit/files/electr onicworkingpapers/ewp18.pdf
42. International Energy Agency (IEA) (2015) World energy outlook
43. Jamal N (2015) Options for the supply of electricity to rural homes in South Africa. J Energy South Afr 3:2413–3051
44. Kambule N, Yessoufou K, Nwulu N (2018) A review and identification of persistent and emerging prepaid electricity meter trends. Energy Sustain Dev 43:173–185
45. Kambule N, Yessoufou K, Nwulu N, Mbohwa C (2019) Exploring the driving factors of prepaid electricity meter rejection in the largest township of South Africa. Energy Policy 124:199–205
46. Kambule N (2015) A survey on the state of energy efficiency adoption and related challenges amongst selected Manufacturing SMMEs in the Booysens area of Johannesburg. Masters minor-dissertation, University of Johannesburg
47. Kelly M (2000) Inequality and crime. Rev Econ Stat 4:530–539
48. Kelly-Detwiler P (2013) Electricity theft: a bigger issue than you think. Forbes. https://www. forbes.com/sites/peterdetwiler/2013/04/23/electricity-theft-a-bigger-issue-than-you-think/# 1e5a20005ed7. Accessed 15 May 2017
49. Khalid SN, Mustafa MW, Shareef H, Aliyu G (2013) Artificial intelligent meter development based on advanced metering infrastructure technology. Renew Sustain Energy Rev 27:191–197
50. Lampietti JA, Banerjee SG, Branczik A (2007) People and power: electricity sector reforms and the poor in Europe and Central Asia. The World Bank, Washington, DC
51. Lier HA (2016) The fourth industrial revolution in the energy industry. https://www.altene rgymag.com/article/2016/02/the-fourth-industrial-revolution-in-the-energy-industry/22789/. Accessed 12 May 2020

52. Lowvelder (2016) Eskom intensifies fight against electricity theft. http://lowvelder.co.za/353 557/eskom-intensifies-the-fight-against-electricity-theft/. Accessed 29 Apr 2020
53. Makanjuola NT, Shoewu O, Akinyemi LA, Ajose Y (2015) Investigating the problems of prepaid metering systems in Nigeria. Pacific J Sci Technol 2:22–31
54. Makonese T, Kimemia DK, Annergarn HJ (2012) Assessment of free basic electricity and use of pre-paid meters in South Africa. http://conferences.ufs.ac.za/dl/Userfiles/Documents/00000/577_eng.pdf. Accessed 13 Oct 2016
55. Martin MW (2014) Pay as you go electricity: the impact of prepay programmes on electricity consumption. Theses and dissertations—agricultural economics, p. 29. http://uknowledge.uky.edu/agecon_etds/29
56. Mauro L, Carmeci G (2007) A poverty trap of crime and unemployment. Rev Dev Econ 11:450–462
57. Mbanjwa T (2017) An analysis of electricity theft: the case study of KwaXimba in eThekwini, KwaZulu-Natal. Masters of Social Science, University of Kwa-Zulu Natal, Published minor dissertation
58. Mburu PT, Sathyamoorthi CR (2014) Switching from post-paid to pre-paid models: customer perception and the organisational role in managing the change: a case study of Botswana power corporation. J Manage Res 3:175–189
59. McRae S (2015) Infrastructure quality and the subsidy trap. Am Econ Rev 105:35–66
60. Miyogo CN, Ondieki NS, Nashappi GN (2013) An assessment of the effect of prepaid service transition in electricity bill payment on KP customers, a survey of Kenya Power, West Kenya Kisumu. Ame Int J Contemp Res 9:88–97
61. Mkhwanazi XH (1999) Electricity as a birthright and the problem of non-payment. In: Paper presented at the third annual South Africa revenue protection conference
62. Musungwini S (2016) A framework for monitoring electricity theft in Zimbabwe using mobile technologies. J Syst Integr 3:54–65
63. Mwaura FM (2012) Adopting electricity prepayment billing system to reduce non-technical energy losses in Uganda: lesson from Rwanda. Utility Policy 23:72–79
64. Nagi J, Yap KS, Tiong SK, Ahmed SK, Mohammad AM (2008) Detection of abnormalities and electricity theft using genetic support vector machines. TENCON 2008—2008 IEEE region 10 conference 19–21 Nov. 2008 Hyderabad, India
65. Navani JP, Sharma NK, Sapra S (2012) Technical and non-technical losses in power system and its economic consequence in Indian economy. Int J Electron Comput Sci Eng 1:757–761
66. News24 (2015) Report illegal electricity theft. http://www.news24.com/SouthAfrica/Local/Maritzburg-Fever/Report-illegal-electricity-theft-20150811. Accessed 13 May 2020
67. North East Group (2020) $96 Billion is lost every year to electricity theft. https://www.prnewswire.com/news-releases/96-billion-is-lost-every-year-to-electricity-theft-300453411.html. Accessed 08 May 2020
68. Onat N (2018) Theft problem and effects of privatization policies on distribution losses of Turkey. J Sci 2:163–176
69. Oseni MO (2015) Assessing the consumers' willingness to adopt a prepayment metering system in Nigeria. Energy Policy 86:154–165
70. Owen G, Ward J (2010) Smart prepayment in Great Britain. Sustainability first, 1–36
71. O'Sullivan KC, Howden-Chapman PL, Fougere G (2011) Making the connection: the relationship between fuel poverty, electricity disconnection, and prepayment metering. Energy Policy 39:733–741
72. O'Sullivan KC, Howden-Chapman PL, Fougere G (2015) Fuel poverty, policy, equity in New Zealand: the promise of prepayment metering. Energy Res Soc Sci 7:99–107
73. O'Sullivan KC, Howden-Chapman PL, Fougere GM, Hales S, Stanley J (2013) Empowered? Examining self-disconnection in a postal survey of electricity prepayment meter consumers in New Zealand. Energy Policy 52:277–287
74. O'Sullivan KC, Viggers HE, Howden-Chapman PL (2014) The influence of electricity prepayment meter use on house hold energy behaviour. Sustain Cities Soc 13:182–191

75. Pretorius, J. 2019: Impact of poor revenue collection and non-technical losses on utilities. https://www.ee.co.za/article/impact-of-revenue-collection-and-non-technical-losses. html. Accessed 16 May 2020
76. Quayson-Dadzie J (2012) Customer perception and acceptability on the use of prepaid metering system in Accra West region of the electricity company of Ghana. http://dspace.knust.edu.gh/bitstream/123456789/4900/1/Quayson-Dadzie%20John.pdf. Accessed 16 Oct 2017
77. Qui L, Xing B (2015) Pre-paid electricity plan and electricity consumption behaviour. http://cmepr.gmu.edu/wp-content/uploads/2014/01/Qiu_pre-paid-pricing_0202.pdf. Accessed 18 Aug 2017
78. Reduan H (2018) Almost 900 electricity theft cases detected in Pahang last year. https://www.nst.com.my/news/nation/2018/01/324432/almost-900-electricity-theft-cases-detected-pahang-last-year. Accessed 13 May 2020
79. Rwanda Energy group (2020) Rwanda Energy Group (REG) condemns theft of electricity as it causes both revenue losses and fatal accidents. http://www.reg.rw/media-center/news-details/news/rwanda-energy-group-reg-condemns-theft-of-electricity-as-it-causes-both-revenue-losses-and-fatal-a/. Accessed 5 May 2020
80. Siebel Energy Institute (2017) Theft detection algorithm thwarts electricity theft from smart meters. http://www.siebelenergyinstitute.org/theft-detection-algorithm-developed-with-support-from-the-siebel-energy-institute-helps-utilities-catch-electricity-thieves/
81. Smart Energy International (SEI) (2017) Analysis: prepaid electricity metering in Africa. https://www.smart-energy.com/features-analysis/analysis-prepaid-electricity-meters-africa/. Accessed 13 May 2020
82. Smith TB (2004) Electricity theft: a comparable analysis. Energy Policy 32:2067–2076
83. Styan JB (2015) Blackout: the Eskom crisis. Jonathan Ball Publishers, Johannesburg & Cape Town
84. Szabo A, Ujhelyi G (2014) Can information reduce non-payment for public utilities? Experimental evidence from South Africa. http://ftp.repec.org/opt/ReDIF/RePEc/hou/wpaper/2014-114-31.pdf
85. Tewari D, Shah T (2003) An assessment of South African prepaid electricity experiment, lessons learned, and their policy implications for developing countries. Energy Policy 31:911–927
86. Timeslive (2015) Soweto shrugs off R4 billion Eskom bill. http://www.timeslive.co.za/local/2015/02/03/Soweto-shrugs-off-R4-billion-Eskom-bill. Accessed 14 Feb 2017
87. Vutete C (2015) Adoption of the prepaid electricity meter billing system by Harare residents: was there some preference to conventional meters? J Bus Manage 9:78–84
88. World Bank (2019) Improving performance of electricity distribution in Brazil. https://www.worldbank.org/en/results/2019/04/24/improving-performance-of-electricity-distribution-in-brazil. Accessed 13 May 2020
89. Yurtseven C (2015) The causes of electricity theft: an econometric analysis of the case of Turkey. Utility Policy 37:70–78

Chapter 4
Prepaid Electricity Meters and Energy Poverty—Lessons from South Africa

Electricity accessibility does not necessarily elicit maximum socio-economic benefits in a society characterised by inequality [3]. This means that "the presence of electricity is by no means a guarantee for development" [57]. Since 1994, South Africa has achieved more than 90% of household electrification rates [48]. The role and contribution of policy in achieving this milestone cannot be understated.

Electricity policies formulated henceforth have essentially, at least theoretically, prioritised maximum and equitable household electrification regardless of household socio-economic positioning and race. Access to electricity has become an 'implied right' in the region [36]. By implication, all households are therefore to be afforded access to basic energy or electricity services for *inter alia* water heating, cooking, and lighting. Hitherto, the household sector consumes more than 30% of the total electricity generated in South Africa [43].

Tied to this marked progress, is the aspect of electricity affordability. This is a crucial feature of socio-economic livelihood and carries cardinal implications on the state of energy poverty in the country [32]. In the past decade, the cost of electricity has increased incessantly. Between 2007 and 2014, the cost of electricity (c/kWh) has risen exponentially by more than 300% [10, 26]. In 2018, the National Energy Regulator (NERSA) approved Eskom's application of an additional 11% tariff increase. Electricity price escalations and expenditure constraints on electricity also tend to push households towards alternative energy sources and voluntarily disconnecting from electricity supply [13, 16, 53].

Scholars have already drawn the link between increasing electricity tariffs, deepened socio-economic vulnerability, and energy vulnerability/poverty among electrified indigent households [32, 44, 45]. Household expenditure or affordability limitation stands as a key indicator for measuring energy poverty status quo [27]. Adam [1] ascertained that in South Africa, between 12 and 20% indigent electrified income is spent on electrical energy. With the increase in electricity tariffs, this expenditure level has certainly increased. Moreover, a study conducted by Winkler et al. [60] also

N. Kambule and N. Nwulu, *The Deployment of Prepaid Electricity Meters in Sub-Saharan Africa*, Lecture Notes in Electrical Engineering 759,
https://doi.org/10.1007/978-3-030-71217-4_4

reports that indigent households connected to the electricity grid in Limpopo, one of the provinces in the country, spend close to 19% of income on electrical energy.

Furthermore, in recent years, Eskom and the household sector have been confronted with respectively unique but intertwined challenging realities associated with the payment for electricity services. For Eskom, it has been the problem of burgeoning household electricity arrearage. In a country of 20 million (close to 50% population size) people considered impoverished, this is however not uncommon. On the other hand, in the past decade, households have been confronted with the challenge of escalating electricity prices—by more than 324% [14]. Poor households therefore remain socio-economically threatened and marginalised. In fact, about 43% of the population is energy poor and the role of increasing electricity prices in this regard cannot be ignored [50]. To manage and reverse the scourge of household arrearage, the utility introduced household prepaid electricity meters (around the 1980s). This was primarily undertaken under the Electricity Basic Services Support Tariff policy (EBSST) (referred to as the Free Basic Electricity (FBE)) in 2003—an elaborate discussion of this is given later.

Several studies and commentaries on the FBE policy or pro-poor incentives have emanated over the years [1, 5, 16, 23, 22, 25, 36, 39, 46, 50, 54, 60]. With the rapidly transforming energy/electricity milieu, there is a broad recognition that there is a need for an updated FBE policy and incentive in order to improve the effectiveness of prepaid electricity meter programme [4, 20, 34, 47, 55]. The effectiveness of the tools is gauged according to the socio-economic value and relevance of its targeted group [4]. For South Africa, as will be seen in the upcoming sections, the FBE remains irrelevant and as such ineffective in dealing with prepaid metered household socio-economic conditions. This then ultimately translates to deepened household energy vulnerability or poverty, also explicated later in the chapter.

This reality has prevailed in South Africa in spite of the existence of several energy policy frameworks and incentives introduced by the government. In the entire region of Sub-Saharan Africa, South Africa's energy regulatory framework is among, if not, the best. It therefore befits to use the country as a reference scenario for understanding how energy poverty can prevail and be averted within an expanding household prepaid electricity market. The lessons drawn are mainly from a Soweto based case study. Soweto, as is the majority of Sub-Saharan African communities, is a predominantly low-income township and therefore broadly representative of prevalent socio-economic conditions across the continent. This chapter is dedicated to show the link between prepaid electricity meters (a technology driving electrification) (discussions on the technology already covered within Chaps. 2, 3 and 4), energy incentives, and energy poverty in South Africa. The lessons on the connectivity of prepaid meters and energy poverty can assist in decision making and general development for the continent.

4.1 Pro-poor Electrification Regulatory and Incentives Framework—Developing Countries

Improving household electrification, particularly in developing countries, is a strategic measure of curbing socio-economic problems and reducing poverty [18, 40]. The electric commodity is a vital development conduit or resource used to meet household basic needs and reduce prevalent socio-economic disparities. Government policies have as a result formulated regulatory energy mechanisms coupled with incentives or programmes that essentially target and seek to benefit poor households [4]. Electricity incentives for poor households, particularly in developing countries, play a specific strategic measure for enabling the affordability of electricity services, improving social equity goals, and bolstering energy policy efficacy [5, 6, 18].

The household tariff structure stands as the most widely used mechanism when designing incentives for the indigent. Authorities adjust tariffs according to the socio-economic conditions, in order to make them attractive and affordable for the poor. Fattouh and El-Katiri [20] point out that low electricity prices assist "the lowest income groups to gain access to modern forms of energy". According to Christensen et al. [12], the sustainability of tariff incentives is dependent on whether the final determined tariff remains relevant to the targeted group, and the ability of the group to afford. The incentive provision of affordable electricity services has thus become central to socio-economic progress in developing countries.

Table 4.2 depicts examples of countries that have formulated and introduced pro-poor household electricity policies and incentives. These largely remain generic and predicated by country-based prevailing socio-economic contexts. Additionally, scholars have shared different perspectives on the role of these incentives [18, 33].

In Sub-Saharan Africa, countries like Ghana and Morocco stand as good examples with pro-poor incentives. In Ghana, the Public Utility Regulatory Commission of Ghana (PURC) runs a social tariff (lifeline tariff) of 50 kWh per month for poor households. There is the availability of imported Compact Fluorescent Light Bulbs (CFL) sold for poor households at highly subsidised rated. The positive impacts of the incentives have however been affected by the increasing electricity tariffs in the country. Morocco serves another specific example with pro-poor incentives. The government disseminates solar photovoltaic home systems at a largely subsidised fee—low-income households receive a 40% subsidy.

Latin American countries have introduced and adopted social tariffs to protect vulnerable households from increasing energy or electricity costs or general economic difficulties [20]. In Peru, low-income households are deemed fit for the social tariff or any social programmes under the SISFOH (Sistema de Focalización de Hogares) system. SISFOH is a central system that collects household socio-economic details and characteristics, then calculates the poverty index and reflects the final determination. In most of these countries, cross-subsidies finance social tariffs, electric companies, or the state. In Argentina, public funds finance incentives. In some countries, customers consuming electricity beyond a certain specified threshold fund the incentives. In the Dominican Republic, the government pays for about 75% of the electricity consumed in the informal settlements [35]. The policy requires that electricity-distributing companies be incentivised, in order to increase the number of consumers included in the programme. Japan has since the 1990s encouraged the use of the DSM program to reduce household electricity consumption. The government and utilities have also applied the Time of Day pricing incentive. Some of the programmes provide educational and employment opportunities to the targeted local communities.

Other research has however contested the purported socio-economic benefit belief of social tariffs and suggested direct welfare payments or investments in social services as more effective means. Fattouh and El-Katiri [20] note that the benefits of energy subsidies tend to leak to high-income households, and therefore recommend that investments should be directed in other social nets (the provision of free public services (e.g. health, education, etc.)) that may guarantee substantive social returns. Because of the leakage, the subsidy may be socially regressive and not beneficial to the targeted group (i.e. indigent households). Banal-Estañol et al. [4] conclude that most incentives in Latin America remain limitedly effective because the incentives are not adapted to the prevalent socio-economic realities. Some of the Organisation for Economic Corporation and Development (OECD) countries have eliminated social tariffs as they engender immaterial change in energy impoverished households. Australia, offers a low-income household rebate of between $285 and $313.50 per year (the noted rates were applicable until June 2018). In 2016, Canada introduced a hydro-electricity rebate for indigent households with an income of less than $50,000. The programme is funded by high-income households whose combined monthly income is more than $50,000. A study by McRae [40] argues social subsidies in Colombia contribute to the tendency of household non-payment and that the state issues them out to expand political constituency and support in order to avoid civil conflict. In the Asian 'Big five' countries (China, India, Japan, South Korea, Indonesia), the subsidy system has proven inefficient [47]. Kemmler [34] and Tongia [55] state that in India, the subsidy benefits middle to high-income households than low-income consumers (the indigent). The economic and social costs of using subsidies in several instances outweigh the intended benefits (Table 4.1) [20].

Table 4.1 Cost of energy subsidies

Type of cost	Description
Economic	Energy subsidies tend: • Reduce incentives for productivity improvements and investments in technology that is more energy efficient • They alter pricing signals resulting in energy waste • The benefits of the incentives tend to leak to high-income households
Social	While they are beneficial to the poor, they are equally regressive because richer consumers tend to benefit from the subsidies
Environmental	They result in increased energy consumption or downplay the importance of energy conservation. This has negative environmental impacts (e.g. increasing airborne emissions and greenhouse gases)

4.2 Pro-poor Electrification Regulatory and Incentives Framework—South Africa

All pre-1994 policies, of the Apartheid system in South Africa, were designed to deepen socio-economic inequality according to race. This translated to socio-economic development framed according to race. Specifically, most black communities, even to date, remain marginalised and indigent [43]. The consequence of the apartheid ideology of 'separate development' was separate budgeting—meaning inter alia underdevelopment of electricity services for black communities. While close to 100% of white suburban areas were electrified, less than 20% of black indigent households were electrified [5, 28].

The emergence of democracy in 1994, under the African National Congress (ANC), sought to reverse this trend and reduce inequality. In the past 24 years, household electrification rates have risen from 36% to more than 90% [10, 48, 51]. The increase has largely been driven by policies and programmes such as the Integrated National Electrification Programme (INEP), Electricity for All, and Reconstruction and Development Programme (RDP) that prioritise equitable socio-economic equity. Among other key regulatory instruments that the government introduced to improve household accessibility and livelihood was the Free Basic Electricity in 2003; incentives included the Demand Side Management (DSM) projects. The next section discusses the noted policy and initiatives.

4.2.1 Electricity Basic Service Support Tariff (Free Basic Electricity) (2003)

In the 1970s, Soweto residents formed a defiance electricity campaign, as an approach of fighting against the Apartheid regime. Through this campaign, households refused to pay their electricity bills. The African National Congress (ANC) promised people *inter alia* free electricity if elected into power. This promise was maintained into the democratic rule. During the 2000 election campaign, free basic water and electricity to residents were at the core of the ANC's local and national government election manifesto. Eskom considers free electricity unrealistic. However, in 2003 the government in collaboration with Eskom introduced the Free Basic Electricity (FBE) Policy. The FBE policy is a complimentary mechanism of the INEP, designed and promulgated by the Department of Minerals and Energy (DME) in 2003. It is thus far the only pro-poor household electricity policy. It centrally seeks to protect vulnerable and impoverished electrified households through bringing relief and ensuring optimum socio-economic benefits from the INEP. The policy serves as a steering tool for the execution of FBE services by municipalities.

Faced with the reality of improving the electrification rate, this instrument was introduced to assuage affordability issues for electrified households. The policy intended to provide basic services to indigent households (grid-tied). The policy asserts that 56% of low-income households have low electricity consumption patterns, and not consuming more than 50 kWh per month. It therefore freely allocates 50 kWh per month (as an incentive) to indigent households and considers this adequate to meet household basic electrical energy needs (i.e. lighting, water heating, and basic cooking or ironing). Just like other developing country pro-poor electricity incentives, similarly this policy instrument seeks to socio-economically protect vulnerable and impoverished electrified households by freely proving them with 50 kWh per month to meet household basic electrical energy needs. One of the fundamental requirements of the 50 kWh electricity provision, is the compulsory installation of a household prepaid electricity meter.

The allocation is concomitant with Eskom's Incline Block Tariff (IBT) beyond the provided 50 kWh. This means that electricity consumption beyond the free allocation is then paid by the consumer. The policy acknowledges that the value of this benefit may be expanded through the application of energy efficiency and conservation initiatives. A Self-targeted (instead of a Broad-Based approach) is used to identify the poor qualifying households. Herein, two approaches exist:

1. Self-targeting approach with Current-Limiting:

   - Household electricity-based needs are limited to a certain (Amperes) current level (e.g. 10 Amp).
   - After exhausting the FBE, the consumer pays the approved tariff per kWh.
   - Households apply with municipality or service provider to be placed under that approved tariff.

2. Self-targeting without Current-Limiting:

   - Households electricity needs are on average or above 150 kWh per month.
   - The exhaustion of the FBE allocation means the consumer purchases electricity at an approved tariff rate per kWh.
   - Consumer applies to be placed under the set approved tariff.

The process of household registration and qualification is administered by the Department of Social Development. The incentive is financed through national government transfers to local government or a cross-subsidy from high electricity consuming customers and richer households (by charging a higher tariff). Household benefits from FBE are both direct and indirect [23, 22, 57]. The direct benefits include monetary savings, protracted electricity use, and appliance purchases. On the other hand, the indirect benefits include improved health, education, environment, and safety, and quality of life.

Currently, more than 51% of households have access to this incentive [50]. Different studies have attached several benefits associated with this incentive, such as monetary savings, extended electricity use, and appliance purchases, improved health, education, environment, and safety and quality of life [5, 23, 22, 25, 39, 46, 57]. However, since 2010 scholars have questioned the socio-economic effect of the incentive for the poor [1, 7, 16, 36, 50, 60]. Some scholars have argued that the incentive contributes little to effect material socio-economic transformation

[5, 39]. Furthermore, the limited rollout of cost-effective measures (e.g. Demand Side Management (DSM) and energy efficient measures) has also fuelled prevalent scholarly assertions that the FBE instrument is obsolete [2, 8, 16, 29, 30, 41, 42, 56, 60]. The role of the DSM project cannot be underestimated, particularly in Sub-Saharan African electrified prepaid metered households that are energy inefficiency, experiencing increasing electricity price, and are most probably already energy vulnerable.

4.2.2 Demand Side Management (DSM)

As is in countries such as Brazil, Mexico, and Latin American countries (Table 4.2), the South African government has recognised the important role of the DSM

Table 4.2 Regulatory and incentives framework for pro-poor electrification programme in developing countries

Country	Electrification rate (%)		Description
Brazil	**97**	• Rural Poverty Alleviation Program (RPAP) (1993)	World Bank grant for financing local grid-connected rural electrification projects. Communities come up with approaches and projects that will be best effective and beneficial to the local conditions
		• Law 10438 (2002)	Lowered tariffs for indigent households
		• *Luz no Campo* (Light in the Countryside) programme (1999) • *Luz para Todos* (LpT) (Light for All) program (2003)	Introduced to improve electricity accessibility and socio-economic development among low-income households. The LpT is the first social-oriented incentive that is cross-subsidised (financed through energy providers) and seeks to reduce social inequality in rural areas. Incentives provided include *inter alia* capacity-building campaigns for local households, education on electricity conservation, security, and efficiency

(continued)

Table 4.2 (continued)

Country	Electrification rate (%)		Description
Other Latin American Countries (Peru, Ecuador, Chile, Dominican Republic, Bolivia, Mexico)	**90**	• Peruvian incentive: FOSE (*Fondo de Compensación Social Eléctrica*—Electricity Social Compensation Fund) (2001) and SISFOH (*Sistema de Focalización de Hogares* • Chile: National Program for Rural Electrification (PER), the National Program on Rural and Social Energy (PERYS), • Mexico: *Luz Sustetable* (Incandescent light bulb replacement programme) (2011)	Promotes accelerated electrification through renewable energy sources Use of social tariffs financed through cross-subsidies. The consumption subsidies differ by country: Peru (0–30 kWh/month); Bolivia (70 kWh/month); Mexico (900 kWh/month—depends on region and season); Ecuador (110–130 kWh/month) In Mexico, there are energy efficiency programmes for light bulbs
India	**85**	• The "Bright Home Programme" (Kutir Jyoti) (1989) • Rajiv Ghandi Grameen Vidyutikaran Yojana (RGGVY) (2005) • Rural Electrification Policy (2006)	The programme was launched to improve electrification among the poor Poor households get free electricity connections with a 100% capital subsidy. Close to 96% of local households qualify for government 50% taxpayer subsidy. For example, all households in Tamil Nadu get free 50 kWh per month Indigent households are to get free electricity connections. But the poor specifically get a 33% (about Rs. 1000) subsidy per annum

(continued)

programme. Its pro-poor policy, which is the FBE, recognises the role energy efficiency initiatives in expediting socio-economic improvement in low-income households. In association with the country's main electricity supplier, Eskom and municipal, the government has prioritised its focus on the DSM programme as a cost-effective measure for improving quality of life. Through the programme and retrofits more than 1 800 MW worth of electricity savings have been generated. This

Table 4.2 (continued)

Country	Electrification rate (%)		Description
Ghana	**83**	• National Electrification programme or scheme (1989) • Self-Help Electrification Programme (SHEP)	An incentive for local engagement wherein the government works with local communities to accelerate the electrification rate. Communities within the 20 km distribution network and have started electrification projects, get support from the government to complete the initiated project earlier than scheduled The Public Utility Regulatory Commission of Ghana (PURC) introduced a social tariff (lifeline tariff) of 50 kWh per month for low-income households Compact Fluorescent Light Bulbs (CFL) have been imported and sold to households at highly subsidised rates. But increasing tariffs have pushed the product prices up again
Morocco	**98**	• Programme for Global Rural Electrification (PERG) (1996)	Dissemination of solar photovoltaic home systems. This is carried out at a fee but largely subsidised by the government. Low-income households receive a 40% subsidy

has transpired through two main initiatives discussed below, namely the Efficient Lighting and Solar Water Heater programme.

4.2.3 Efficient Lighting Programme (2007)

Another demand management mechanism employed by Eskom and the Department of Energy (DoE) is towards reducing household electricity demand was the distribution of Compact Fluorescent Lights (CFL). The country launched this cost-effective initiative in 2004, and the goal was to freely distribute 43.5 million energy efficient

CFLs by 2010. This was a door-to-door exchange programme involving the replacement of high consuming incandescent light bulbs with efficient CFLs. By 2014 more than 60 million light bulbs had been deployed. About 30,000 persons were employed from this project.

4.2.4 Solar Water Heater Programme (2008)

One of the direct outcomes of the White Paper on Renewable Energy (2003) was the Solar Water Heater Programme. In 2008, the Eskom conceptualised rebate or subsidy programme, firstly named Geyser Load Reduction Programme. The aim of the programme then was to reduce electricity demand and consumption, particularly among the high-end electricity consumers. The later was in the light of the electricity crises. In 2009, the programme was initiated by the DoE and became known as the National Solar Water Heater programme. The target was to have 50% of South Africa's household water heating through solar water heating technology by 2020. This was going to be achieved by deploying 1 million solar water heaters by 2015. The plan was that, between 2009 and 2012, the programme would be funded through the National Treasury. But this never effectively materialised because of the financial crises experienced by Eskom in 2013 and the programme was halted. The social component of the programme aimed at delivering hot water services through low-pressure solar water heater systems to indigent households across 54 municipalities. Close to 90,000 systems have been distributed thus far. Rebates are only offered to households with high-pressure systems. The installations are cost-free. By 2017, only 30,000 systems had been installed for indigent households. Although a negligible number of households have the systems installed, the put government has recently budgeted R411 million to intensify the roll-out exercise.

Another important incentive given in South Africa, as done in other developing countries, is the social tariff. The tariff is characterised by electrification grants, the already mentioned FBE, and cross-subsidies whereby the richer are charged more to subsidise households at the lower end of the income distribution.

Unfortunately, to date, none of these programmes have been sustainable—the effectiveness is questionable either because of infrastructural, administrative, or management issues. In the light of increasing tariffs and deepening energy poverty, it is important to design programmes or introduce mechanisms that will ensure optimal programme efficacy. The next section opens up with an elaborate discussion on energy poverty in the context of a prepaid electricity metered market. More importantly, mechanisms to deal with the challenge are then proposed.

4.3 Energy Poverty

Energy poverty only occurs where households use electricity or gas [11]. Fuel poverty on the other hand places traditional sources of energy (wood, etc.) at the centre. Energy poverty is generally associated with low-income households. While the concept of energy poverty is generic and the definition varies by geography, but it is broadly recognised the poverty expenditure threshold standard at 10–15% of household income [9, 44, 27]. This is household income spent to meet energy related needs such as cooking, lighting, heating and cooling. These services are necessary for human development and the lack of access thereof poses a socio-economic threat to the households. One of the contributing factors for households to struggle in accessing the services, and rendering them energy vulnerable, is the increasing cost of electricity.

4.3.1 Increasing Electricity Tariffs

One of the determinants of energy poverty in South Africa is access to affordable electricity [27]. This is one of the central imperatives of FBE policy. In the past decade, electricity tariffs have increased by more than 324% [10, 14]. The incessant increase has been attributed to the need for Eskom to strengthen its financial basis for infrastructure and electricity capacity building, in order to avert another load shedding incidence. This surge in cost has however affected the household's ability to access affordable and sufficient electricity and has rendered households vulnerable and further impoverished. Households that have access to electricity through the national electrical grid have become more prone to energy poverty than those off the grid [27].

South Africa uses the expenditure approach to measure energy poverty. For example, a household earning a monthly income of R10 000 ($654.5) (assumed rate of R15.2/$), will have to spend between R1 000 ($65.4) and R1 500 ($98.1) on energy to be considered energy poor [32]. To date, about 43% of South Africa's households are classified as energy poor. Energy poverty is not unique to South Africa. Regions such as the United Kingdom, Northern Ireland, and New Zealand have 20, 44, and 23% energy poverty levels, respectively [24]. International studies, for instance by O'Sullivan et al. [44] and Chester [11], have already confirmed the link between increasing tariffs and energy poverty. The rising tariffs may translate in a surge of disconnection rates because of the non-payment of bills, resulting in deepened energy poverty and further strained household budget [24]. The reality that is facing the FBE policy is that increasing electricity tariffs reduce the value of the 50 kWh incentive and therefore having immaterial socio-economic impact among poor households.

4.3.2   Energy Efficiency

The FBE policy acknowledges the value of cost-effective measures such as energy efficiency in reducing energy poverty. One of the realities that remain unchanged in most grid-connected low-income households of South Africa is that they continue to depend on old inefficient electric appliances for heating, cooking, and refrigeration need [36]. This is while the energy efficiency market has expanded over the years. For instance, current energy efficient refrigerators use at least 75% less energy than a ten year old refrigerator model. Progressive as the market has been, but the upfront cost of energy efficient home appliances remains to be relatively high for indigent households to affordably access in South Africa. Take for example, the basic differences between the incandescent, Compact Fluorescent Lightbulbs (CFL), and Light Emitting Diodes (LED) lightbulbs.

Energy efficient LED lightbulbs are beneficial in their average life span, stretching to a duration period of 25,000 h or 10 years. Associated with this may be the environmental benefits. Poor households, however, choose appliances almost exclusively based on price and brand, not on duration [31, 29, 38, 52]. Therefore, given the socio-economic reality and the light-bulb prices, the likelihood of poor households choosing a R10.00 ($0.64) energy inefficient incandescent lightbulb as compared to an energy efficient one with a ten-fold higher cost, is higher.

After more than a decade of the promulgation of the DME [17], energy efficiency remains unaffordable for poor households. Whereas the government has established the DSM programme (e.g. Efficiency Lighting and Solar Water Geyser project) targeting poor households, in the course of its deployment and operation it has been plagued with challenges and as such been halted [2, 8, 16, 41, 42, 60]. With all this, the programme's socio-economic effectiveness is therefore questionable. With household energy inefficiency being another indicator of energy poverty [44], this prevalent condition denotes that the FBE policy has failed in assisting household maximise the benefit of the 50 kWh incentive. Low-income households therefore remain trapped in the cycle of energy poverty.

The arguments expressed thus far border on the fact that increasing electricity prices and the pervasive energy inefficiency in low-income households of South Africa render the FBE policy obsolete. The next discussion unfolds in relation of this stance and primarily extends it by arguing, as opposed to mainstream ideology, that in the context of FBE policy, prepaid electricity meters for poor households are socio-economically ineffective.

4.4   Prepaid Electricity Meters and Energy Poverty

The utilisation of prepaid electricity meters is deemed as another mechanism that government and Eskom have proclaimed as a means of improving quality of life for impoverished households. Additionally the technology was introduced to manage and mitigate the problem of household non-payment for electricity services. In the past decade, the government has intensified the deployment of the technology. Hitherto, more than 66% (4.2 million) of the residential sector in the country is prepaid metered [29]. The goal is to have all households prepaid metered by 2020 [29].

While over the years the installation of the technology was based on consumer willingness, in 2016 Eskom declared that the installations should be on a 'compulsory-basis for all households' to *inter alia* curb the utility's losses from household non-payment [21]. In 2003, Eskom in an effort to encourage its customers to pay for the electricity consumed, concurrently promulgated the FBE policy to make electricity affordable and erased R1.4 billion for one township's debt. Regardless, by 2015 the debt had escalated again—this time six-fold (R8.6 billion) [49]. Herein has been another motivation for deploying the technology, as a credit management tool. Some of the overall pronounced benefits of the technology include: electricity and monetary savings therefore reduce energy poverty, the elimination of stress associated with bills, and useful tool to aid budgeting [19, 37].

A plethora of the aforementioned studies and the FBE policy fundamentally fail to acknowledge that household prepaid electricity meters may elicit different socio-economic outcomes for different income households [29]. Meaning that, the purported benefits of the technology may not be true for all household income groups. In South Africa specifically, the rollout of prepaid electricity meters under the current FBE status quo, will socio-economically marginalise low income households.

In an analysis of prepaid meter based electricity consumption data (2007–2014), for 3 841 low-income households of Chiawelo—the first area in Soweto township to receive the technology—it was ascertained that although prepaid electricity meters resulted in 48% electricity reduction (Figs. 4.1 and 4.2), because of prevalent socio-economic realities this finding should be interpreted with caution [32]. This is because it was further proven that a low-income household in the area that earns on average R992 per month[1] spends about 66% of their monthly income on prepaid meter based electricity (Table 4.3), which ultimately renders this household income group energy impoverished (i.e. ratio: income vs. electricity). So, while Eskom is arguably recovering the non-payment debt, this is however coming at a socio-economic detriment for poor households.

Poor prepaid metered households will either spend more in order to avoid disconnection or self-disconnect because they cannot afford the electricity. With most of these households located in urban locations, alternative fuel and energy sources such as wood for cooking and water heating may be inaccessible. So, while prepaid electricity meters result is reduced electricity consumption, but they have a strong potential for further entrenching energy poverty. In a prepaid metered household context, the government's 50 kWh only contributes about 7% to the final prepaid based electricity consumption of an indigent household [32]. This is inadequate to meet the basic energy needs stated in the FBE policy of lighting, water and space heating, basic cooking and ironing. In view of this reality, the FBE policy is recognised as obsolete as it fails to achieve its set goal of reducing energy poverty through improving access to affordable prepaid meter electricity services.

---

[1]Based on StatsSA [48] data.

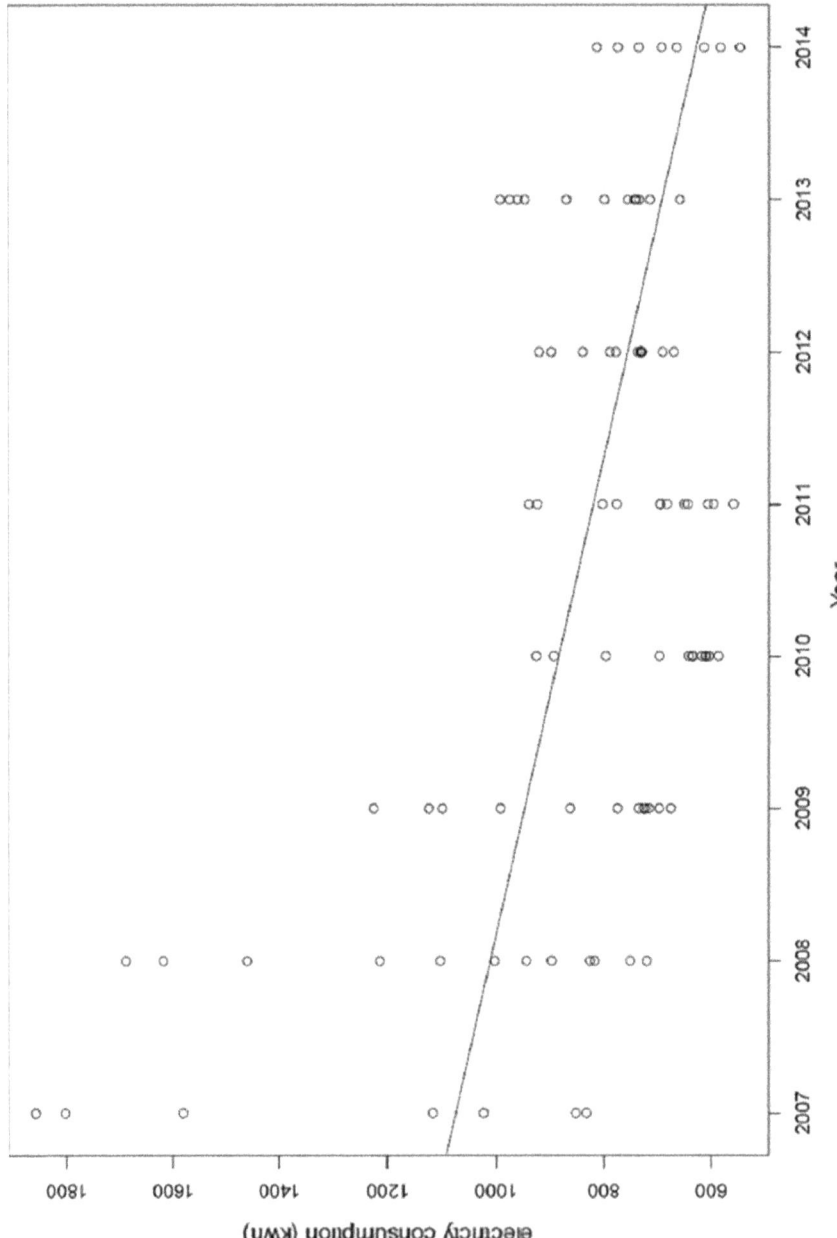

Fig. 4.1 Electricity consumption trend among prepaid metered low-income households of Chiawelo, Soweto (2007–2014) [32]

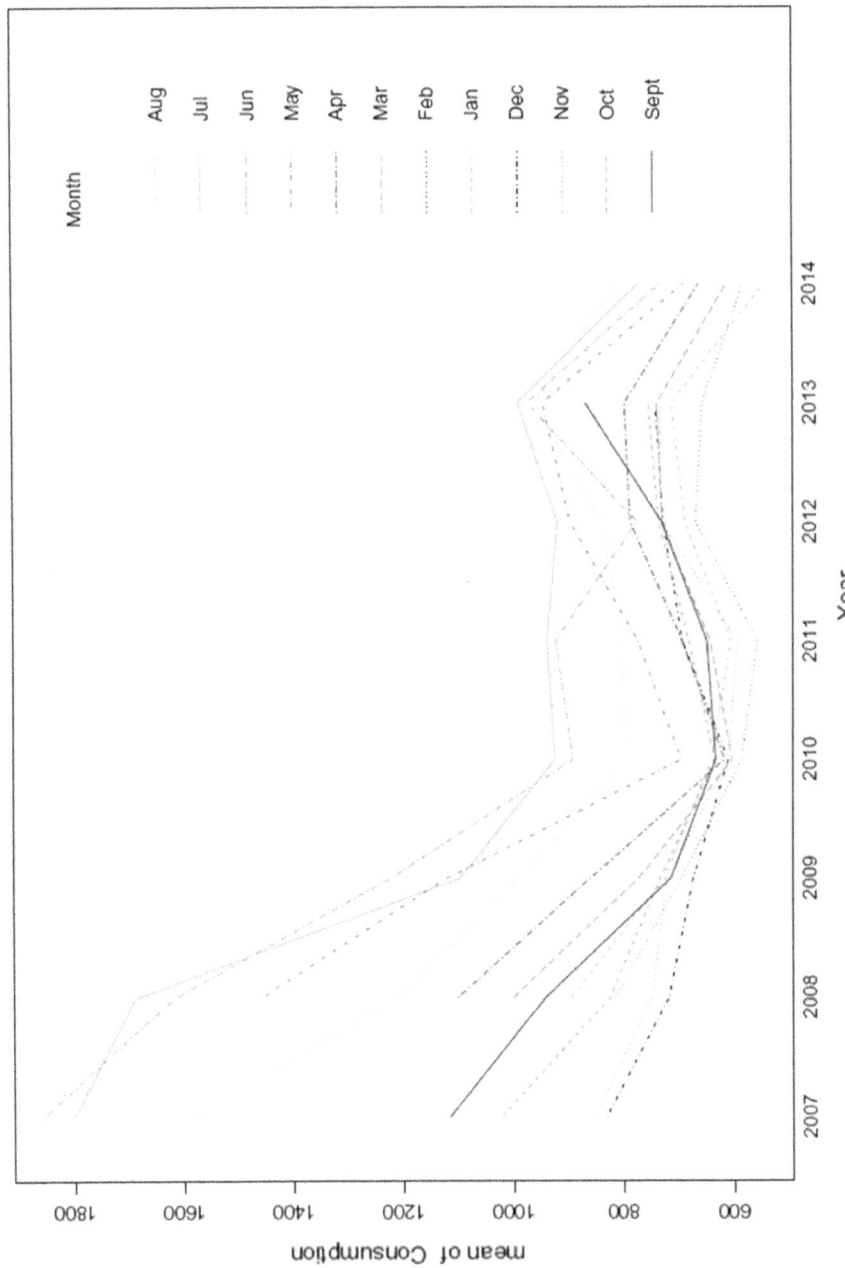

Fig. 4.2 Annual electricity consumption pattern (monthly trend) for low-income households (2007–2014) [32]

Table 4.3 Prepaid meter based electricity consumption and expenditure for an average low-income household [32]

Prepaid meter based electricity consumption per month (Based Eskom's 2014 data for 3 841 low income households; this is the overall total for consumption (including space and water heating and etc.))	Assumed tariff	Electricity expenditure per month
667.6 kWh	R0.98 per kWh	R642.2 ($43.8) per month

An effective prepaid electricity meter programme or market for poor households will require a consideration of several aspects. There is a need to reengineer the prevalent social contracts based on the current and emerging socio-economic status quo. In the past 15 years, the FBE policy remains unchanged and proves to be socio-economically irrelevant [1, 27, 36]. One key factor that has rendered the policy instrument irrelevant or ineffective is the aforementioned rise in electricity costs in the past decade and it remaining unchanged. While the 50 kWh incentive, in and of itself has been considered immaterial (Dlamini 2016) [43], the escalating price of electricity has further reduced the value of the social value of the incentive. Despite that, no efforts have thus far been put to review and update the state of the FBE instrument. Research has inferred that the efficacy and sustainability of a tariff incentive is dependent on its relevance to the target population [12]. The specific policy areas that need to be reviewed to improve the relevance of the instrument include the size of allocation and Self-targeting approach with Current-Limiting (i.e. 20 A). Household electricity consumption needs, use exceedingly more (60 amp) than the Current-Limiting stipulated in the policy document [1, 36].

Scholars also generally concur on the need for relevant pro-poor household electricity incentives, particularly in developing regions such as Sub-Saharan Africa (Matsukawa et al. 2000; Elinder et al. 2016) [4, 5, 12, 35, 40, 47]. The FBE policy, for example, stands as one of the pro-poor electricity policies and incentives that need to be revised. Since its promulgation, no review has been undertaken—in spite of increasing electricity prices. One of the additional components of the revised document is an the establishment of a policy monitoring and review committee. The policy review process should be made mandatory and carried after a set number of years. This will ensure that the tool is timeously improved to be constantly relevant to the socio-economic conditions prevailing among indigent households. A context-specific approach is therefore a prerequisite.

Moreover, the current prepaid electricity meter programme should be redesigned by 2020. A new design should thus feature a process characterised by a context-specific deployment-link of prepaid meters and cost-effective measures (e.g. energy efficiency). The aim is to concomitantly increase the level of energy efficiency and deploy prepaid meters in a manner that will elicit maximum socio-economic benefits, and ultimately reduce the 66% household expenditure levels on prepaid meter electricity. Energy efficiency measures are the cornerstone for prepaid electricity meter efficacy [29]. These will assuage the impact of increasing electricity prices, which are likely to continue increasing into the future. The role of the established FBE monitoring and review committee will include managing the proposed deployment-link. This is because the effectiveness of the FBE social incentive will be framed by the extent of energy efficiency among prepaid metered low-income households.

In addition to the context-specific approach, a holistic method has been suggested by Kambule [30]. This method is comprised of 10 benchmark factors that should be considered if a prepaid meter programme is to be effective and thus reduce energy poverty in low-income household settings. The factors are tabulated and described in Table 4.4.

For electrification to be fully effective, it has to be tied with incentive programmes and strategies that will render the electricity affordable and reduce energy poverty. Up to now, this is not the case. Increasing tariffs for low-income households are further entrenching energy poverty [44, 45]. White et al. [58] remarked that electricity at the time remained an expensive commodity for poor households. This status quo persists in spite of the existence of FBE policy and other noted pro-poor incentives. For that reason, studies have in recent years recommended alternative approaches for improving the effectiveness of the pro-poor electricity incentives in South Africa. For example, Adam [1] has argued for an expanded allocation of 200 kWh (poverty incentive); Bhorat et al. [7] suggests a reconsideration of the household poverty line (<R1 500), to be raised beyond this level; improved social education and energy efficiency programmes [27]; use of carbon tax to reduce energy poverty [59].

Overall, research studies advocate that the following be undertaken in order to improve the socio-economic value and relevance of pro-poor electricity incentives in South Africa:

- A need to re-evaluation current pro-poor electricity instrument, particularly its nature of design
- Strengthened focus on efficient and sustainable DSM incentives to reduce the affordability burden. Technologies that prevent communities from developing should be not be imposed upon the poor [60]
- Household based capacity building for optimal effectiveness of the policy and other incentives.

While Eskom and government are applying noticeable effort in improving poor households' socio-economic living, by promulgating pro-poor electricity incentives and programmes, thus far most of these efforts are largely ineffective. The current prepaid electricity meter programme is an example of such an initiative. The deployment process currently does not consider the differences in socio-economic conditions prevalent among different households. All poor households qualifying for the FBE 50 kWh incentive are required to install prepaid electricity meters as a means of reducing energy poverty.

To date, households were noted to spend about 66% of their monthly income on electricity. A context-specific approach that is relevant to the socio-economic realities of poor households is as such recommended to reduce the energy poverty levels. Escalating electricity prices and energy inefficient poor households also render the

Table 4.4 Benchmark factors characterizing a holistic approach to deploying prepaid electricity meters [30]

Benchmark	Description
Target group	The introduction of the prepaid meter programme in 2007 was arguably supposed to reduce electricity debt due to household electricity non-payment. However, the target has now shifted to mainly covering all newly built government houses for the poor
Non-payment	Electricity non-payment is a major issue in Soweto, particularly amongst households that are Eskom's clients. By the end of 2015, the township owed Eskom close to R9 billion, and 86% of the households did not pay their electricity bills. So, the technology is a credit management tool for non-paying customers
Disconnection/Reconnection	Prepaid electricity meters installed in Soweto disconnect the consumer immediately after the credits are exhausted. Reconnection is only activated when the meter is recharged
Marketing, education, and training	In introducing prepaid meters marketing, training, and education become central aspects for familiarising the target population with the technology
Stakeholder (community) involvement	The involvement of all key stakeholders in the design of the prepaid meter programme and its roll-out thereof is important
Electricity tariff and consumption	The tariff market is regulated by National Electricity Regulator of South Africa (NERSA). All households, with or without a prepaid electricity meter are subject to similar tariffs, as approved by NERSA. Furthermore, the installation of prepaid electricity meters will have a bearing on household electricity consumption
Incentives	The prepaid electricity meter programme for poor households in South Africa comes with a free 50 kWh monthly incentive—based on the FBE Policy (2003). Besides this electricity subsidy, there exists none
Appliance/Energy efficiency	Electricity appliance and their nature of efficiency play an important role in the consumption of electricity, and therefore the effect of prepaid meters in households. A majority of low-income households in Soweto use high consuming appliances, leading to the credit finishing faster than a household using an energy efficient appliance. Moreover, the size and age of dwelling are energy inefficient, making the households energy vulnerable
Regulator tools	A widespread use of prepaid meters may require a need for regulatory (codes of practice) guidelines dealing with the governance of the programme. Currently, Eskom has no prepaid electricity meter guidelines but has voiced out the importance of having this

(continued)

Table 4.4 (continued)

Benchmark	Description
Leadership and management	The electricity sector history of South Africa is political. Political leaders encouraged township residents to boycott payments—enhancing the culture of non-payment. It may require equal leadership to encourage people to pay and adopt prepaid meters

FBE policy obsolete because these realities reduce the socio-effectiveness of the 50 kWh incentive and consequently also deepen energy poverty. The policy framework is to be reviewed and constantly updated based on prevailing socio-economic realities. Secondly, the deployment of prepaid electricity meters should be concomitant with that of cost-effective measures aligned with DSM. This is so that the level of household expenditure on prepaid meter based electricity may be reduced. The proposed approach will ensure that while Eskom recovers the debt owed, poor households are also sustainably benefiting from prepaid electricity meters. South Africa is a global pioneer in household prepaid electricity meters, perspectives shared alternative approaches improve policy and programme effectiveness may be reciprocated in other Sub-Saharan African countries that aim to expand their prepaid meters market and equally reduce energy poverty for low-income households.

Lastly, no Sub-Saharan African state, including South Africa has a specific prepaid electricity meter regulatory mechanism that governs the deployment and functionality of the prepaid meters programme. In the light of the expanding market and the noted implications on household socio-economic conditions, thus energy poverty, it is important that this be considered by the individual regional countries.

4.5 Highlights

- *Coherent and up to date policy framework*: The presents of a regulatory framework is pivotal for giving guidance to the expanding prepaid electricity meter market. Additionally, these regulations need to be clear and up to date. This means that they have to be timeously aligned to the socio-economic conditions or needs of the households.
- *Prepaid electricity meter regulatory mechanism*: Given the expanding market and the impact that the technology may have on household livelihood, it is important to formulate laws to protect households from *inter alia* energy poverty. This will ensure that both the authorities (for management) and households (payment of electricity services without being affected) are accountable.
- *DSM project or incentives*: No prepaid electricity meter programme should run without a DSM project or incentives. Energy efficiency programmes should be an integral component of prepaid meter programmes. Moreover, poor households should be given incentives to optimise the benefits of prepaid electricity meters.

- *Context-specific and holistic approach*: Before the deployment of the technology, authorities should dedicate time to assess the prepaid meter target market, in order to create a programme that context-specific. This means that the programme will be relevant to the prevalent socio-economic conditions on the ground and therefore can reduce energy poverty in the community. This can be done, for example, by considering the benchmark factors suggested by Kambule [30].

References

1. Adam F (2010) Free basic electricity: a better life for all. Johannesburg, Earthlife
2. Aigbavboa (2015) Low-income housing residents' challenges with their government install solar water heaters: a case of South Africa. Energy Proc 75:495–501
3. Aliu IR (2020) Energy efficiency in postpaid-prepaid metered homes: Analysing effects of socio-economic, housing, and metering factors in Lagos. Energy Effic, Nigeria. https://doi.org/10.1007/s12053-020-09850-y
4. Banal-Estañol A, Calzada J, Jordana J (2017) How to achieve full electrification: lessons from Latin America. Energy Policy 108:55–69
5. Bekker B, Eberhard A, Gaunt T, Marquard A (2008) South Africa's rapid electrification programme: policy, institutional, planning, financing and technical innovations. Energy Policy 36:3125–3137
6. Bezerra SPBG, Callegari CL, Ribas A, Lucena AFP, Portugal-Pereira J, Koberle A, Szklo A, Schaeffer A (2017) The power of light: socio-economic and environmental implications of a rural electrification program in Brazil. Environ Res Lett 12:
7. Bhorat H, Oosthuizen M, van der Westhuizen C (2012) Estimating a poverty line: an application to free basic municipal services in South Africa. Development Policy Research Unit, University of Cape Town
8. Blommestein KC, Daim TU (2013) Residential energy efficient device adoption in South Africa. Sustain Energy Technol Assess 1:13–27
9. Bouzarovski S, Petrova S (2015) A global perspective on domestic energy deprivation: overcoming the energy poverty–fuel poverty binary. Energy Res Soc Sci 10:31–40
10. Chehore T (2014) South Africa: electricity pricing paralysing the poor. http://www.ee.co.za/article/south-africa-electricity-pricing-paralysing-poor.html
11. Chester L (2014) Energy impoverishment: addressing capitalism's new driver of inequality. J Econ Issues 2:395–404
12. Christensen JM, Mackenzie GA, Pedersen MB (2012) Enhansing access to electricity for clean and efficient energy services in Africa
13. Clark A, Davis M, Eberhard A, Gratwick K, Wamukonya N (2005) Power sector reform in Africa: assessing the impact on poor people. University of Cape Town
14. Deloitte (2017) An overview of electricity consumption and pricing in South Africa. An analysis of the historical trends and policies, key issues and outlook in 2017. http://www.eskom.co.za/Documents/EcoOverviewElectricitySA-2017.pdf. Accessed 18 Jan 2018
15. Department of Minerals and Energy (DME) (2003) Electricity basic services support tariff (free basic electricity) policy. http://www.energy.gov.za/files/policies/Free%20Basic%20Electricity%20Policy%202003.pdf. Accessed 12 Mar 2018
16. Dlamini KTH (2015) Coping mechanisms of low-income urban households to escalating energy costs in South Africa. Masters dissertation, University of Witwatersrand
17. DME (2005) Energy Efficiency Strategy of the Republic of South Africa. http://www.energy.gov.za/files/esources/electricity/ee_strategy_05.pdf. Accessed 13 Apr 2013.

18. Elindera M, Escobara S, Petrea I (2017) Consequences of a price incentive on free riding and electric energy consumption. PNAS 12:3091–3096
19. Esteves GRT, Oliveira FLC, Antunes CH, Souza RC (2016) An overview of electricity prepayment experiences and the Brazilian new regulatory framework. Renew Sustain Energy Rev 54:704–722
20. Fattouh B, El-Katiri L (2012) Energy subsidies in the Arab world. Research paper series bassam.fattouh@oxfordenergy.org
21. Fin24 (2016) Molefe moots compulsory prepaid electricity for all. Available Online http://www.fin24.com/Economy/Eskom/molefe-compulsory-prepaid-power-will-curb-eskom-losses-20160921
22. Gaunt CT (2005) Meeting electrification's social objectives in South Africa, and implications for developing countries. Energy Policy 33:1309–1317
23. Gaunt CT (2003) Researching a basic electricity support tariff in South Africa. Domestic use of energy conference
24. Gonz´alez-Eguino M (2015) Energy poverty: an overview. Renew Sustain Energy Rev 47:377–385
25. Howells M, Victor DG, Gaunt T, Elias RJ, Alfstad T (2006) Beyond free electricity: the costs of electric cooking in poor households and a market-friendly alternative. Energy Policy 34:3351–3358
26. Inglesi-Lotz R, Blignaut JN (2011) Estimating the price elasticity of demand for electricity by sector in South Africa. South Afr J Econ Manage Sci 14:449–465
27. Ismael Z, Khembo P (2015) Determinants of energy poverty in South Africa. J Energy South Af 3:66–78
28. Jack BK, Smith G (2015) Pay as you go: pre-paid metering and electricity expenditure in South Africa. https://sites.tufts.edu/kjack/files/2015/08/Jack_manuscript7.pdf. Accessed 15 Nov 2016
29. Kambule N, Yessoufou K, Nwulu N (2018) A review and identification of persistent and emerging prepaid electricity meter trends. Energy Sustain Dev 43:173–185
30. Kambule N, Yessoufou K, Nwulu N, Mbohwa C (2019) Exploring the driving factors of prepaid electricity meter rejection in the largest township of South Africa. Energy Policy 124:199–205
31. Kambule N (2015) A survey on the state of energy efficiency adoption and related challenges amongst selected manufacturing SMMEs in the Booysens area of Johannesburg. Masters minor-dissertation, University of Johannesburg
32. Kambule N, Yessoufou K, Nwulu N, Mbohwa C (2018b) Temporal analysis of electricity consumption for in prepaid- metered low- and high- income households in Soweto, South Africa. Af J Sci Technol Innov Dev. https://doi.org/10.1080/20421338.2018.1527983
33. Katz J, Kitzing J, Schröder ST, Andersen FM, Morthorst PE, Stryg M (2018) Household electricity consumers' incentive to choose dynamic pricing under different taxation schemes. Energy Environ 7:1 of 14
34. Kemmler A (2007) Factors influencing household access to electricity in India. Energy Sustain Dev 4:13–20
35. Krishnaswamy V, Stuggins G (2007) Closing the electricity supply-demand gap. World Bank Energy Min. Sect. Board Discuss, 20
36. Makonese T, Kimemia DK, Annergarn HJ (2012) Assessment of free basic electricity and use of pre-paid meters in South Africa. http://conferences.ufs.ac.za/dl/Userfiles/Documents/00000/577_eng.pdf. Accessed 13 Oct 2018
37. Malama A, Mudenda P, Ng'ombe A, Makashini L, Abanda H (2014) The effects of the introduction of prepayment meters on the energy usage behaviour of different housing consumer groups in Kitwe, Zambia. AIMS Energy 3:237–259
38. Manuel AL, Flores Q (2006) Study of an ordinary home electrical consumption in order to make it more energy-efficient. Renew Energy Power Quality J 1:1–8
39. Marquard A, Bekker B, Eberhard A, Gaunt T (2007) South Africa's electrification programme an overview and assessment. Working paper
40. McRae S (2015) Infrastructure quality and the subsidy trap. Am Econ Rev 105:35–66

41. Monyei CG, Adewumi AO (2017) Demand side management potentials for mitigating energy poverty in South Africa. Energy Policy 111:298–311
42. Monyei CG, Adewumi AO (2018) Integration of demand side and supply side energy management resources for optimal scheduling of demand response loads—South Africa in focus. Electr Power Syst Res 158:92–104
43. Moshoeu LB (2017) Critical analysis of the right to access electricity for the destitute in South Africa: issues and challenges. Masters dissertation, University of Limpopo
44. O'Sullivan KC, Howden-Chapman PL, Fougere G (2011) Making the connection: the relationship between fuel poverty, electricity disconnection, and prepayment metering. Energy Policy 39:733–741
45. O'Sullivan KC, Howden-Chapman PL, Fougere G (2015) Fuel poverty, policy, equity in New Zealand: the promise of prepayment metering. Energy Res Soc Sci 7:99–107
46. Prasad G, Visagie E (2006) Impact of energy reforms on the poor in Southern Africa—explicit focus on the poor. Energy Research Centre, University of Cape Town
47. Rehman IH, Kar A, Banerjee M, Kumar P, Shardulm M (2012) Understanding the political economy and key drivers of energy access in addressing national energy priorities and policies. Energy Policy 47:27–37
48. StatsSA (2018) Community survey 2016: Statistical release P0301. www.statssa.gov.za
49. Styan J (2015) Blackout: the Eskom crisis. Jonathan Ball Publishers, Johannesburg & Cape Town
50. Sustainable Energy Africa (2017) Energy poverty and gender in Urban South Africa. www.sustainable.org.za
51. Tewari D, Shah T (2003) An assessment of South African prepaid electricity experiment, lessons learned, and their policy implications for developing countries. Energy Policy 31:911–927
52. Tholen L, Götz T, Covary T, Thomas S, Adisorn T (2016) Harnessing appliance energy efficiency in South Africa: Policy gaps and recommendations to address actor-specific barriers
53. Thopil GA, Pouris A (2015) Aggregation and internalisation of electricity externalities in South Africa. Energy 82:501–511
54. Tinto EM, Banda KG (2005) The integrated national electrification programme and political democracy. J Energy South Af 4:26–33
55. Tongia R (2017) Delhi's household electricity subsidies: highly generous but inefficient? Brookling India Impact series No. 042017
56. Uken PE (2012) Pitfalls of SWH. Energy Proc 30:1432–1434
57. Wentzel M (2005) Achieving universal access to electricity in South Africa. Energize, 12–15
58. White C, Bank L, Jones S, Mehlwana M (1997) Restricted electricity use among poor urban households. Dev South Afr 3:413–423. https://doi.org/10.1080/03768359708439974
59. Winkler H (2017) Reducing energy poverty through carbon tax revenues in South Africa. J Energy South Afr 3:12–26
60. Winkler H, Simoes AF, Rovere EL, Alam M, Rahman A (2011) Access and affordability of electricity in developing countries. World Dev 6:1037–1050

Part II
Sub-Saharan Region: Adoption and Experiences

Chapter 5
Western Africa Region

5.1 Ghana

5.1.1 Geographical Overview

The Republic of Ghana is a country in West Africa on the Gulf of Guinea with geographical coordinates 8°00'N and 2°00'W, total area of 238 533 km^2 and sharing borders with Burkina Faso, Côte d'Ivoire and Togo. The Republic of Ghana, shown in Fig. 5.1, has the world largest artificial lake called Lake Volta [19]. According to the World Bank [72, 73], Ghana's population as of 2019 is 30 417 856. However, Worldometer [75, 76] estimates that the population of Ghana as of November, 2020 is 31,327,040 with a population density of 137 per km^2. Geographically, Ghana is flat with coastal savannahs, tropical rain forests and sandy beaches with her capital city named Accra [21]. The Republic of Ghana is the first African country to gain independence in 1957 with English Language as her official language in addition to several national indigenous languages. The Ghanaian people are adherents of Christianity, Islam and Traditional faiths [16].

Economically, Ghana has a rich natural resource base of gold, cocoa and oil along with industries in mining, manufacturing, lumbering, food processing, smelting, aluminium, cement and small commercial ship building [19]. The GDP of Ghana as of 2019 is $65.3 billion and a GDP per capita of $2,164 [30].

5.1.2 Energy Overview

Energy is the fulcrum of any economy and a primary driver for prosperity and increase in living standards, especially for a developing economy. Ghana's energy resources comprise of oil deposits, natural gas reserves, coal reserves and renewables such

© The Author(s), under exclusive license to Springer Nature Switzerland AG 2021   79
N. Kambule and N. Nwulu, *The Deployment of Prepaid Electricity Meters in Sub-Saharan Africa*, Lecture Notes in Electrical Engineering 759,
https://doi.org/10.1007/978-3-030-71217-4_5

Fig. 5.1   The Republic of Ghana [16]

as wind and solar energy [6, 19]. Electric energy for the Republic of Ghana comes from thermal energy, hydro-electricity, natural gas, solar, diesel, and imports from Côte d'Ivoire [6, 35]. According to the country's energy ministry, the total installed capacity for electricity generation is 4,132 MW, where hydro sources makes up 38%, thermal sources 61% and solar energy sources makes up less than 1%. According to the 2019 Electricity Supply Plan for the Ghana Power System [33], generation plants are from the government owned Volta River Authority (VRA) and independent power producers (IPPs) as shown in Table 5.1.

The generated electricity is transmitted through the Grid Company (GRIDCO), distributed by two government owned distribution companies (Electricity Company of Ghana, ECG and Northern Electricity Department, NED) and one privately owned distribution company named Enclave Power Company, EPC Ltd, [35]. The demand for electricity in Ghana ranges from 3500 MW and is projected to rise to 4,500 MW by 2024 as shown in Fig. 5.2 [33]. Additionally, it is estimated that the energy consumption for 2019 was 17,237.79 GWh including transmission network losses (898.03 GWh) [33]. However, according to Our World in Data [58], the per capita electricity consumption in Ghana as of 2019 is 471 kWh, as shown in Fig. 5.3. Moreover, access to electricity in Ghana is at 85% with an electrification rate of 93% in urban areas and 75% in rural areas, with the population without access to

Table 5.1 Ghana's generation plants [33]

Plants	Installed capacity (MW)	Dependable capacity (MW)	Fuel type
Akosombo G	1 020	900	Hydro
Kpong GS	160	105	Hydro
Tapco (T1)	330	300	LCO/GAS
TICO (T2)	340	320	LCO/GAS
TTIPP	110	100	LCO/GAS
TT2PP	80	70	GAS
KTPP	220	200	GAS/DIESEL
VRA Solar Plant	2.5	0	Solar
TOTAL VRA	**2,263**	**1,995**	
BUI GS	404	360	Hydro
CENIT	110	100	LCO/GAS
AMERI	250	230	GAS
SAPP 161	200	180	GAS
SAPP 330	360	340	LCO/GAS
KAR POWER	470	450	HFO
AKSA	370	350	HFO
BXC	20	0	SOLAR
MEINERGY SOLAR	20	0	SOLAR
TROJAN	44	39.6	DIESEL/GAS
GENSER	22	18	GAS
CEN POWER	360	340	LCO/GAS
AMANDI	2,820	2,598	
TOTAL IPP	**2,820**	**2,598**	
TOTAL (VRA + IPP)	**5,083**	**4,593**	

electricity given as 5 Million [40]. Moreover, electricity, which is the dominant form of modern energy in Ghana accounts for 65% of energy in Ghana's industrial and services sector and about 36% in the residential sector [25].

In the management of Ghana's energy sector and electricity sub-sector, below are the relevant stakeholders [31, 32]:

a. Ministry of Energy: this is the ministry responsible for energy policy formulation, monitoring, implementation and coordination of all the activities relating to energy in Ghana both electricity and petroleum sub—sectors [35].
b. Electricity Company of Ghana (ECG) Ltd: this is a limited liability company owned by the state, which distributes and sells electricity to southern Ghana, which comprises of Ashanti, Central Greater Accra, Eastern and Volta Regions of Ghana, making it the largest distribution company in Ghana [39].

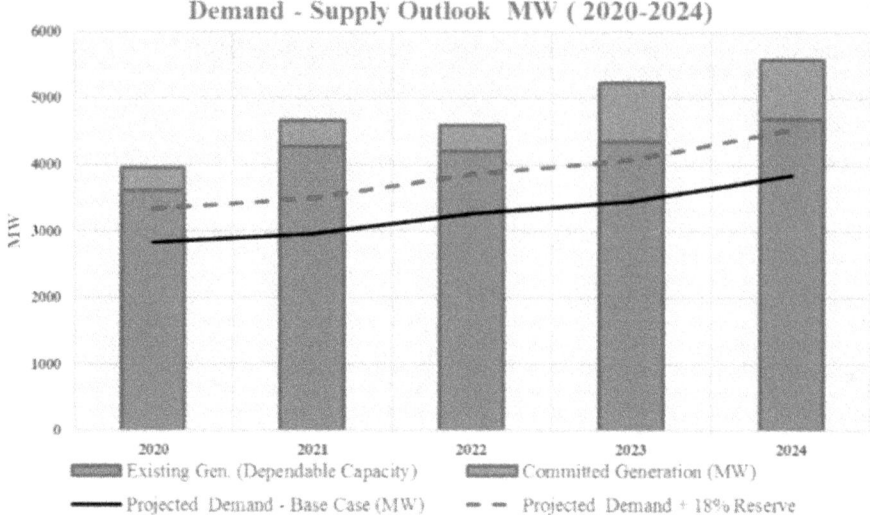

Fig. 5.2   Ghana electricity demand and supply outlook (2020–2024) [33]

Ghana: **Per capita**: how much electricity does the average person consume?

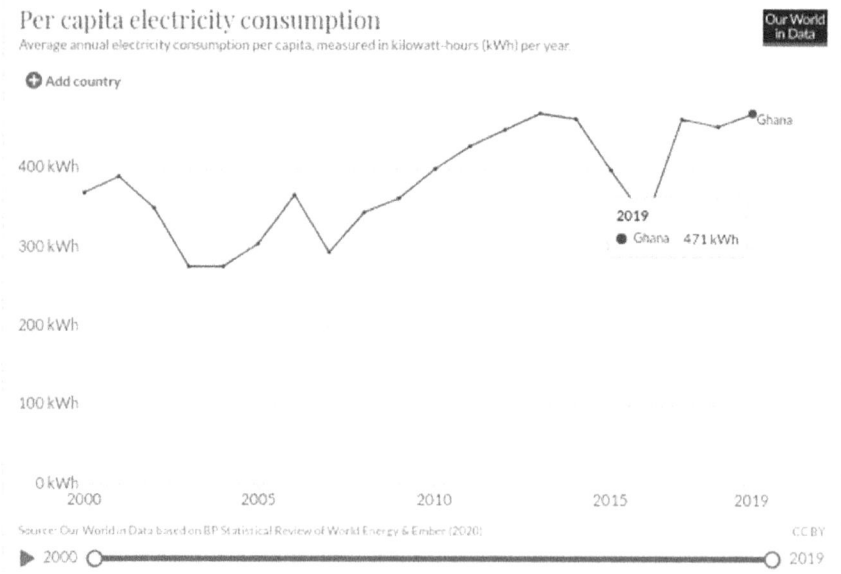

Fig. 5.3   Ghana's electricity consumption per capita [58]

c. Ghana's Grid Company (GRIDCO): this is the state electricity transmission company, which transmits electricity from electricity generating companies to customers such as Electricity Company of Ghana (ECG), Northern Electricity Company (NEDCo) and the mines [37]. The company also provides metering and billing services and manages assets required to transmit energy.

d. Volta River Authority: this is the company responsible for the generation of electricity in Ghana, currently controlling eight [8] power plants in Ghana and working in Renewable Energy (RE) policy [71].

e. Northern Electricity Distribution Company: this is the distribution company owned by the state responsible for distributing electricity to Northern Ghana which comprises of Berong–Ahafo, Northern, Upper East and Upper West regions of Ghana [35].

f. Energy Commission: this is Ghana's electricity/energy regulator and advisor to government on energy matters on the regulation, utilization, development and management of energy resources in Ghana [26].

g. Enclave Power company Ltd (EPC): This is a private owned electricity distribution company in Ghana, covering Tema free zones enclave in the Greater Accra Region, with 50 industrial customers in the Region [35].

h. The Public Utilities Regulatory Commission (PURC): this is a government agency responsible for the regulation and provision of guidelines for electricity rates and rates chargeable for utility services [60].

Furthermore, below are the laws, regulations and policies guiding Ghana's electrical energy sub-sector [26, 36, 43]:

- Energy Sector Levies (Amendment Act, 2019)
- Petroleum (Exploration & Production) General Regulation 2018
- Energy Commission Act, 1997 (Act 541)
- Public Utilities Regulation Commission Act, 1997 (Act 538)
- The Volta River Development Act 1961 (Act 46)
- Bui Power Authority Act 2007 (Act 740)
- The Renewable Energy Act 2011 (Act 832)
- National Electricity Grid Code (The Grid Code) 2009
- The Environmental Protection Agency Act, 1994 (Act 490)
- Electricity commission (Technical, operations, and standard of performance) Rules, (2008)
- The Atomic Energy Commission Act (2000)
- National Energy Policy 2010 (The Policy)
- Electricity Supply and Distribution (technical and operational) Rules, 2005
- Electrical Wiring Regulations 2011
- Electricity Supply and Distribution (Standards of performance) Regulations, 2008
- Renewable Energy Act (2001)

5.1.3 Prepaid Electricity History and Current Status

According to Azila et al. [14] and Boadu [15], the Electricity Company of Ghana (ECG) whilst attempting to solve issues such as inefficient cash flows, elimination of bad debts, ineffectual revenue collection and ensure financial probity adopted the prepaid metering system. The prepaid metering system started as a pilot test in 1994 and 1995 using 'cash power' in areas such as Sakumono Accra, Kumasi and Tema for both commercial and residential consumers[14].

The prepayment meter is an electronic instrument used for the supply of electricity and premeasurement of the amount of power a customer consumes [15, 38]. According to the authors, a prepayment system is an advance purchase of electricity to be consumed and the electricity supply stops after the expiration of the purchase rate of electricity. Furthermore, UK Power Limited [69] opined that the prepayment meter works like its name and does not entail receiving payment for electricity consumed at the end of the month, because the bills are already paid before consuming the electricity.

According to Boadu [15] and O'Sullivan et al. [51], the following are the benefits of using prepaid meters in Ghana:

• A decrease in billing and disconnecting customers.
• Significant improvement in generation of revenue, which also leads to decrease in working capital for the distribution companies.
• Understanding how energy consumption takes place leading to proper control of energy use and proper budgeting.
• Reduces the practice of customers delaying in paying overdue bills.
• It ensures transparency in its operations requiring no deposits.

In Tuffour et al. [66] five types of prepaid meters are used in Ghana and they include; Smart cash, Pay and smile, Electro—cash, cash power and BOT/BXC. According to Quayson and Dadzie [62], there are different categories of prepaid meters, and they are summarized as follows:

a.   Integrated Single Phase (ISP) electric meter: it is a compact keypad based prepaid electricity meter typically used in households compatible with standard display information such as low credit warning, load contractor status and energy consumption using the Liquid Crystal Display (LCD). The meter, which is keypad oriented, supports Standard Transfer Specification (STS) and supports the algorithms of the 20—digit STS encryption.

b.   The Integrated Three Phase (ITP) Meter: this is a 4—wire 100 Amp per phase, keypad based meter, which is suitable for residential, commercial and light industrial environments. It has diagnostic indicator features showing the status of communication to the remote Customer Interface Unit (CIU). The meter comprises of critical metering, token decryption and load control functionality.

c.   The Split Single Phase (SSP) meter: this is a two wire, keypad—based prepayment electricity meter with two parts, namely the Energy Management Unit

(EMU) and the Customer Interface Unit (CIU). Also, meter information such as low credit warning, energy consumption and load contactor status are displayed on the CIU.

According to Tuffour et al. [66], prepaid meters were introduced in Ghana for two main reasons:

- To address ECG operational challenges such as ensuring efficacy in the utilization of electricity, reduction in ECG's operational cost, high cost of customer billing by ECG, tampering of post-paid meters by customers to evade payment, delay in payment of bills and high indebtedness by corporate bodies and private individuals.
- To solve squabbles and conflicts between landlords and tenants over bills payment.

The 1994/1995 pilot test of prepaid meters in Ghana led to it's ful launch in 2005, extending to major cities, metropolis, municipal and district capitals, with the installed meters produced by Ghana Electrometer [14]. As of 2013, about 30% of ECG customers have purchased and are using prepaid meters [23]. Furthermore, according to Ghana's electrometer company, it has distributed 440,000 prepaid meters as of 2014 [44]. Also, in 2019, ECG installed more than 4,000 prepaid meters in Kibi city of Ghana [34]. Although there are no available updates on the number of prepaid meters currently deployed in Ghana, ECG's website [24] shows the type of prepaid meters used and their deployment areas as summarized in Table 5.2. Therefore, logically comparing Table 5.2, the 2014 installation numbers and the 2019 Kibi installation with World Population Review [74] of cities in Ghana and their population, it could be adduced that over 60% of ECG customers and electricity consumer have prepaid meters across Ghana and about 800,000 prepaid meters have been installed.

5.1.4 Challenges of Prepaid Metering

Despite the benefits and advantages of prepaid meters as introduced in Ghana, there have been various consumer complaints about the technology. For instance, prepaid consumers complain that their electricity expenses are higher than postpaid consumers. Alfred [9] stated that the use of prepaid metering is quite expensive due to Ghana being a 'developing country' and the fact that the prepaid meter runs faster than post-paid, meaning more money is needed to buy more units. The author further stated that the use of prepaid metering in Ghana is riddled with faulty meters and the difficulty in inputting or uploading the units coupled with the expiry dates of cards. Another challenge enumerated by the author is the nonchalant behaviour of workers responsible for installing, repairing and responding to the need of the users.

Quayson and Dadzie [62] stated that the use of prepaid meter has benefitted the distribution companies especially the Electricity Company of Ghana (ECG) over and above Ghanaians. The author further opined that the adoption of prepaid meters has

Table 5.2 Types of prepaid meters used in Ghana [24]

Prepaid Type	Features	Areas of Deployment
E-Cash I	• It has a smart card for payment	• Cape coast • Kasoa • Kasoa south • Swedru • Akim Tafo • Koforidua • Nkawkaw • Ho • Agona • Bibiani • Takoradi • Afienya • Krobo • North Tema • Nungua • Prampram • South Tema
E-Cash II	• It has a user interface unit • It has a smart card	• Dodowa • Kwabenya • Legon • Makola • Mampong • Roman Ridge • Teshie
E-Cash III	• It has a smart meter but not operated as such • It has a RFID card • It has a contact less smart card	• Agona • Axim • Bibiani • Half Assini • Takoradi • Sekondi • Ho • Hoshoe • Afienya • Krobo • North Tema • Prampram • Nungua • South Tema • Akim Tafo • Koforidua • Cape coast • Kasoa • Kasoa south • Swedru • Winneba • Dodowa • Kwabenya • Legon • Makolo • Roman Ridge • Teshie

(continued)

Table 5.2 (continued)

ENER SMART (Holley)	• It uses smart GPRS Technology • It has a user interface unit • It has a RFID card	• Ashanti East PDS Operational region
KAMSTRUP (NURI)	• It uses smart ZIGB and GPRS technology • Customer gets meter information via SMS	• Dodowa, Kwabenya • Legon • Makola • Akuapim – Mampong • Roman Ridge • Teshie • Krobo • Prampram
SMART G	• It has a RFID card • It uses contactless smart card • It has KEYPAD user interface unit • Uses GPRS technology to read data from meters.	• Danyame
SMARTCASH—BOT	• It is not a smart meter • It has a RFID card • Its uses contactless smart card/Remote card for loading money onto the meter from vending point	• Kaneshie • Bortianor • Nsawam • Achimota • Dansoman • Korle – Bu
SMARTCASH—BXC	• It uses a PLC and GPRS smart technology. • It has a user interface unit • It has a RFID card • It uses a contactless smart card for loading money into the meter from the vending point	• Accra West (Bortianor) • Accra East (Teshie) • Tema (Nungua)

reduced in Ghana due to the high cost of prepaid meters in the country. Moreover, Tuffor et al. [66] noted that the disadvantages of prepaid meters in Ghana are technical faults in meters, the expensive nature of the prepaid meters, difficulty in uploading prepaid units, delays in obtaining a meter, scarcity of prepaid units, and low voltage.

Furthermore, Asante [13] noted that despite the benefits of prepaid meters, there are several challenges which includes; long & winding queues at private vending points and ECG revenue collection centre, whenever there is an internet disruption leading to revenue loss. The authors also stated that prepaid meters users in Ghana experience voltage problems whenever electricity is cut off at the slightest voltage hikes or drop in voltage levels, which doesn't affect post-paid customers.

5.1.5 Future Prognosis

The traditional power grid in Ghana utilizes one directional power flows and is yet to fully embrace a key technological breakthrough in the energy/electricity industry, which is the advent of the Smart Grid [12]. The traditional grid system in Ghana has suffered from blackouts, lack of robust outage management systems, vulnerability to tampering activities and inevitable human errors.

Smart Grid are intelligent grids that integrate a variety of distributed energy resources (DER), smart meters, smart appliances, energy efficient resources, advanced control of power distribution, voltage and frequency synchronizers with improved monitoring devices [1]. Armah [12] opined that smart grid is a transformation from a centralized producer—controlled networks to one that is not centralized and enables remote monitoring and customer- interaction whilst remaining efficient, reliable, flexible, and grid visibility through the use of latest information technologies.

It is therefore obvious that smart meters, a key component of the smart grid is a necessary evolution in the Ghanaian prepaid metering sector. ECG has pioneered this in 2014 [59] in partnership with GRIDCo and DNVGL. This initiative should be further deployed and widespread in order to take full advantage of technological breakthroughs in the energy industry i.e. smart metering [1, 2, 12, 77].

5.2   Nigeria

5.2.1   Geographical Overview

Nigeria is a country located in West Africa with geographical coordinates of 10°00 N and 8°00E. Niger, Chad, Cameroon, Republic of Benin, Gulf of Guinea and Equatorial Guinea are Nigeria's neighbours as shown in Fig. 5.4 [20, 67]. According to Worldometer [75, 76], Nigeria's population as of November 2020 is 208,076,483, making the nation the 7th largest country in the world and Africa's most populous country with a population density of 226 per km^2 and a total land area of 923 777 km^2. Ubogu [67] stated that Nigeria's land area spans across six ecological zones, from swampy coastal rainforest of the south, semi-arid fertile grassland of the eastern region, northern region & western region coupled with creeks & lagoons of the Niger Delta and drainage system of the Niger–Benue Rivers.

The official language of Nigeria is English and there are three major indigenous languages in the country, namely Hausa, Yoruba and Igbo. The country also has three [3] major religions; Christianity, Islam and Traditional religions [19, 67]. Nigeria is a culturally diverse country with different ethnic groups spanning across 36 autonomous states further divided into 774 local governments and the Federal Capital Territory, Abuja [72, 73].

Economically, Nigeria is the largest economy in Sub-Saharan Africa with a GDP of 475 billion USD as of 2019, annual growth of 2.3% and a GDP per capita of 2,363 USD [29]. The Nigeria economy is highly dependent on oil accounting for 10% of the country's GDP, 70% of government revenue and 83% of the country's total export earnings [49].

Fig. 5.4 The Federal Republic of Nigeria (Britannica 2020b)

5.2.2 Energy Overview

Energy is an indispensable and crucial factor for the growth of a nation's economy as it affects other sectors like agriculture, education, commerce and manufacturing [17, 50]. Nigeria has huge energy resources comprising of fossil fuels (Nigeria is the sixth largest producer of crude oil in the world and possesses 193.35 trillion cubic feet of gas), ample deposits of coal, wood fuel, wind energy, tidal energy and hydropower energy [55].

Nigeria has an installed electricity generation capacity of 12,522 MW mainly from thermal sources (10,142 MW) and Hydro sources (2,380 MW), however the quantum of electric energy available for use by consumers ranges from 3,500 to 5,000 MW (USAID, 2020) [31, 32]. Moreover, as shown in Fig. 5.5, the NERC [47] report stated that as of the first quarter of 2020, available generated electricity is 3,912 MW consisting of a 73.45% thermal source and 26.55% hydro sources. In addition, the 2019 electric consumption per capita in Nigeria was 152 kWh [64] with 62% of the population having access to electricity and 77 million Nigerians have no access to electricity (predominantly in the rural areas [40]. In addition, Nigeria's current supply of electrical energy only meets one—third of its demand for electricity [61].

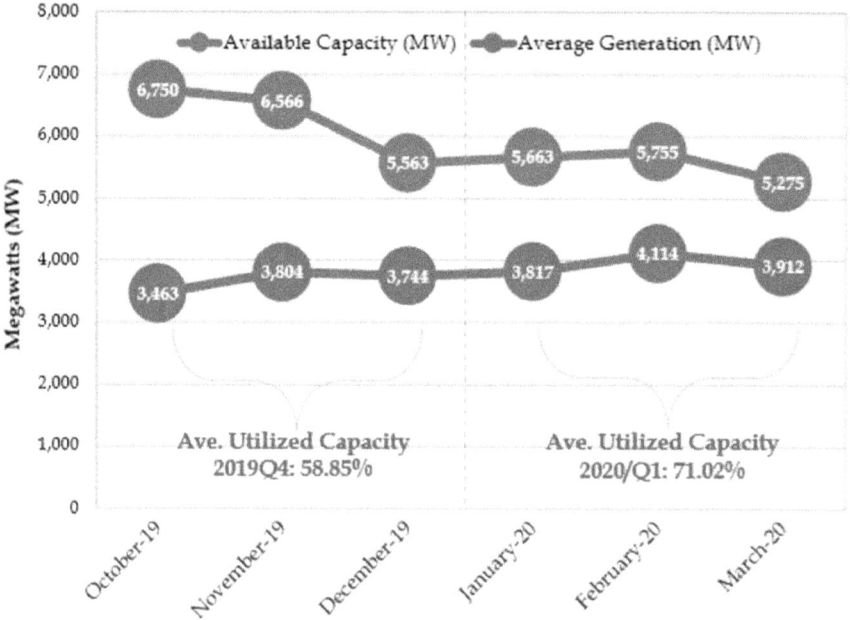

Fig. 5.5 Nigeria's electricity generated as of first quarter of 2020 [47]

Historically, the Nigerian electric energy sector, otherwise known as the power sector has its roots in the Electricity Corporation of Nigeria (ECN), established in 1950. Subsequently in 1962, the Niger Dams Authority (NDA) was established for the construction and maintenance of dams for electricity generation. The NDA was instrumental to the construction of the Kainji dam which was commissioned in 1968 [45, 55]. ECN and NDA were merged to create the National Electric Power Authority (NEPA) in 1972 responsible for the generation, transmission and distribution of electricity. Subsequently NEPA metamorphosed to Power Holding Company of Nigeria (PHCN) in 2005 [28, 55].

The privatization process of NEPA, which led to PHCN, further led to the unbundling of PHCN leading to six (6) electricity generation companies, eleven (11) electricity distribution companies and one power transmission company, succinctly described below [27, 55]:

a. Generation companies: through the privatization of 2005, the government divested its stake in its thermal power plants and hydropower stations which led to six [6] generation companies, otherwise known as GENCos, listed in Table 5.3:

Table 5.3 Electricity generating companies of Nigeria [55]

S/N	GENCO	Type	Installed capacity	Privatization status
1.	AFAM VI power Plc	Thermal	776 MW	100% sold
2.	Sapele power Plc	Thermal	414 MW	51% sold
3.	Egbin power Plc	Thermal	1 020 MW	100% sold
4.	Ughelli	Thermal	900 MW	100% sold
5.	Kainji/Jebba power Plc	Hydro	1338 MW	Longterm concession
6.	Shiroro hydro electric Plc	Hydro	60 0 MW	Longterm concession

The generated electricity is sold to the Nigeria Bulk Electricity Trading (NBET) at an agreed price based upon the power purchase agreements, which then collaborates with the transmission company of Nigeria (TCN).

b. Transmission Company of Nigeria (TCN): this is a company fully owned by the Federal Government and managed by the private sector from the unbundling of PHCN in 2005. It is the link between generation and distribution of electricity and oversees the grid system, reduces system failures and ensures that sectors players are in full compliance with the National grid code. The TCN has three departments namely; Transmission Service Providers (TSP), System Operation (SO) and Marketing Operation.

c. Distribution companies (DISCOs): there are 11 distribution companies over a coverage area closer to the consumers. Table 5.4 shows the eleven DISCOs and their areas of operation:

In the Nigeria Power sector, apart from the GENCOs, TCN and DISCOs, other stakeholders are:

a. Nigerian Electricity Regulatory Commission (NERC): this is an independent regulatory body that promotes and ensures the power sector is investor friendly and the market structure is efficient to meet the needs of Nigeria Electric Demand.

b. Nigeria Bulk Electricity Trading (NBET) Plc: this is the administrative agency that manages Nigeria's electricity pool in the market, acting as the bulk purchaser of electric energy in Nigeria electricity supply industry.

Moreover, regulating laws, policies, and regulations guiding the industry includes the following [55, 68]:

Table 5.4 Electricity Distribution companies of Nigeria [47, 55]

S/N	DISCOs name	Area of coverage
1.	Abuja electricity distribution company Plc	Abuja, Kogi, Nasarawa and most parts of Niger state
2.	Benin electricity distribution company Plc	Edo, Ondo, Delta and parts of Ekiti state
3.	Eko electricity distribution company	Southern Lagos: Ojo, Festac, Ijora, Mushin—orile, Apapa, Lekki, Lagos Island and Agbara in Ogun state.
4.	Jos electricity distribution company Plc	Bauchi, Benue, Gombe and Plateau state
5.	Kaduna electricity distribution company Plc	Kaduna, Kebbi, Sokoto and Zamfara state
6.	Kano electricity company Plc	Kano, Kastina & Jigawa state
7.	Ibadan electricity distribution company Plc	Oyo, Ogun, Osun, Kwara and Parts of Niger state, Ekiti and Kogi state
8.	Ikeja electricity distribution company Plc	Parts of Lagos state: Ebule Egba, Akowonjo, Ikeja, Ikorodu, Oshodi, Shomolu
9.	Enugu electricity distribution company Plc	Abia, Anambra, Enugu, Ebonyi and Imo state
10.	Port–Harcourt electricity distribution company Plc	Akwa—Ibom, Bayelsa, Cross-River and River state
11.	Yola electricity distribution company Plc	Adamawa, Borno, Taraba and Yobe State

a. Electric Power Sector Reform Act (EPSR) of 2005
b. National Electric Power Policy Document of 2001
c. Regulations for the Investment in Electricity Network, 2015
d. Nigerian Electricity Supply and Installation Standards Regulations 2015
e. Regulations on National Content Development for the Power Sector 2014
f. NERC Regulation for the Procurement of Generation Capacity 2014
g. NERC (Embedded Generation) Regulation 2012
h. NERC (Independent Electricity Distribution Networks) Regulations 2012
i. NERC Permit for Captive Power Generation Regulations 2008
j. NERC Mini Grid Regulation 2016
k. Nigerian Electricity Smart Metering Regulation 2015
l. Feed in Tariff for Renewable Energy Sourced Electricity in Nigeria 2015
m. NERC Meter Asset Provider (MAP) regulation 2018
n. NERC Eligible Customer Regulations 2017
o. The Market Rules 2009
p. Grid Connection Policies
q. National Renewable Energy and Energy Efficiency Policy
r. National Renewable Energy Action Plans
s. Rural Electrification Strategy and Implementation Plan
t. National Mass Metering Program (NMMP) Implementation 2020

u.   Competitive Transition Change Regulations
v.   Capping of Estimated Billing Regulations.

5.2.3 Prepaid Electricity History and Current Status

In combating the issue of over—billing of customers for electricity and other defects associated with electricity consumption, metering was introduced in Nigeria in 2005 [45]. Simpson [63] defined metering as the method and procedure by which devices are used in measuring the amount and direction of energy flow, especially by the end-user. Moreover, according to Kettless [42], prepaid metering is a system where a customer pays for energy before using it. Also, Ajenikoko and Adelusi [7], Nextier [48] and Malama et al. [46] opine that prepaid electricity/energy meters ensure that energy/electricity bills are paid for by end- users or customers prior to its consumption. Moreover, Fagbohun and Femi–Jemilohun [28] further reiterated that due to the disadvantages of post–paid metering and the customer debt profile resulting from inadequate revenue collection, the then Power Holding Company of Nigeria, PHCN in 2005 introduced the pre-paid system, which entails the purchase of electricity credit in customers electricity account before usage. The prepaid meter is like using airtime on a prepaid mobile line, where a customer controls what he uses [3].

Amhenrior et al. [10] detailed the various energy prepaid meter models in Nigeria, summarized in Table 5.5.

From the research of Ajenikoko and Adelusi [7], Adekitan et al. [4], Fagbohun and Femi–Jemilohun [28], Ogbuefi et al. [53]; the benefits and advantages of introducing prepaid metering systems in Nigeria include:

a.   Eliminating issues of unpaid bills and inaccurate bills.
b.   Up-front payment for electricity.

Table 5.5 Types of prepaid meters models in Nigeria [10]

S/N	Name	Type	Compact/Split	Keypad type
1.	Actaris	Token	Compact	Compact wired
2.	Itron	Token	Split	Wired
3.	A. Conlog (Bec 23)	Token	Compact	Compact wired
	B. Conlog (Bec 44)	Token	Split	wired
	C. Conlog (Bec 44)	Token	Split—wireless	Wireless (Rf based—150 m)
4.	A. Mojec	Token	Split	wired
	B. Mojec	Token	Split	Plc
	B. Mojec	Token	Split	PLC
	B. Mojec	Token	Split	
5.	Mommas	Token	Split	Wired

c. It enables efficient consumer energy budgeting in accordance with their financial position and lifestyle.
d. It helps in gauging utilization and monitoring of electricity units because the remaining credits are displayed.
e. There is reduction in the level of contact between energy provider officials and consumers.
f. It improves cash flow and revenue management for DISCOs.
g. The DISCOs benefits from reduction in cost associated with billing notification of disconnection and reconnection, customer service staff and call centres.
h. It improves detection and management of outages.
i. The prepaid meter will help and enable the distribution companies to determine their revenue.

However, despite the significance and importance of prepaid meters to both the consumer and distribution companies, there is still a large number of customers without meters in Nigeria [52]. Historically, the prepaid meter was introduced as a revenue generating system in 2005 by PHCN to collect bills from evading customers [57]. The lack of a widespread deployment of prepaid meters in the power sector led the NERC in 2012 to introduce a meter intervention scheme called "Credited advance payment for metering implementation (CAPMI)", where consumers will self–finance the meter cost and repay over a period at a 12% interest rate per annum [48]. After the privatization of PHCN in November, 2013, DISCOs planned a metering target of supplying 4.92 million Meters within three years with an annual target of 1.640 million meters, which they failed to achieve [3]. In 2016, according to the data from NERC, about 45% of 7.47 Million customers nationwide possessed prepaid meters [48]. According to Adams [3], available data from NERC shows that as of December 2018, only 3.7 million customers have been metered out of over 8 million. Collaborating this, Orient Energy Review [56] citing the NERC website, shows that out of above 8 million registered active electricity customers, only 44.6% have been metered.

These figures consistently show that over 55% of registered customers with various DISCOs are yet to be metered from 2006 till 2020 [5]. Moreover, the metering status of DISCOS as of March, 2020, as released by NERC [47] quarterly report, shows that there's an increase in registered customers and a slow progress of metered customers. Furthermore, according to NERC, the current metering gap as of October 2020 is over 10 million, comprising of unmetered customers and customers with obsolete meters that need to be replaced [18]. However, in the report of Nigeria's widely read and recognized newspaper, Vanguards Newspaper [70], in its report quoting a metering expert stated that Nigeria needs 24.7 million prepaid meters, with majority in the real estate sector of the economy (Fig. 5.6).

In meeting this metering gap, NERC approved the meter asset providers (MAP) regulations in 2018, with the objectives of allowing independent meter providers in agreement with Distribution companies to supply prepaid meters to the Nigerian Electricity Consumers [27]. As cited by the author, the objectives of MAP include [27]:

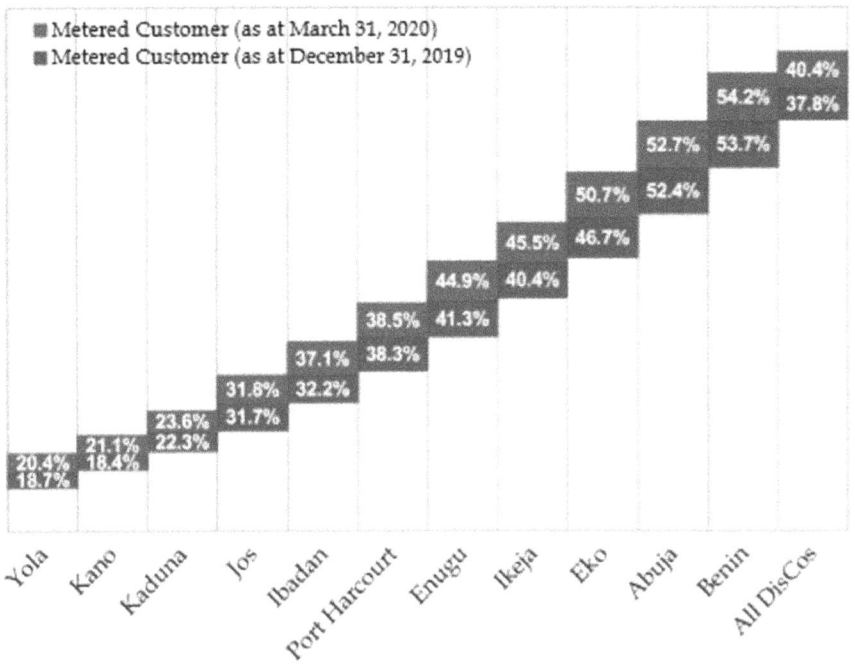

Fig. 5.6 Metering status of DISCOs customers as of first quarter 2020 [47]

a. Encourage the development of independent and competitive metering services
b. Eliminating estimated billing practices
c. Attracting private investment to the provisions of metering services.
d. Closing the metering gap through accelerated meter roll out.
e. Enhancing revenue assurance in the Nigerian Electricity Supply Industry (NESI).

5.2.4 Challenges of Prepaid Metering in Nigeria

In Makanjuola, et al. [45] and Orient Energy Review [56]; the authors x-rayed the challenges and problems associated with prepaid metering system in Nigeria, which are succinctly explained below:

a. Lack of vending infrastructure in some locations: the authors argued that vending machines helps in the generation of prepaid tokens and helps conclude associated prepayment transaction. This machine is lacking in some areas/locations for prepaid meters deployment, hence some DISCOs are unable to deploy prepaid meters because of the symbiotic relationship between vending machine and prepaid meters.

b. High cost of purchasing a prepaid meter: prepaid meters cost as much as thrice the amount of post-paid meters. As of 2015, a single-phase prepaid meter was ₦39, 375 and three phase prepaid meter was ₦68, 901. However as of 2020, the NERC increased the price of prepaid meter for single phase to ₦44, 896.16 and three phase to ₦82, 855.19, citing reasons such as hike in foreign exchange rate for the increment.

c. Absence of Local Manufacturers: there is no presence of competent local manufacturers in Nigeria, thus delaying the mass procurement of meters for Nigerians as envisaged by the Federal Government and in the purchase agreement with DISCOs.

d. Lack of Expertise: when prepaid meters have fault and gets damaged, there is inadequate expertise in the country to fix the faulty/damaged meters culminating in meter abandonment and long waiting period to export for mass repair.

e. Unavailability of service on Sundays and Holidays: the customer service oriented prepaid metering stops on Saturday and is unavailable on Sundays and during public holidays irrespective of when their electricity unit finishes.

f. Single phase overloading: due to ignorance, many customers overload a single phase prepaid meter beyond its capacity, which thus leads to its damage.

g. Delay in getting and installing prepaid meters.

Furthermore, Adams [3], Orukpe and Agbontaen [57] opined that the eleven [11] distribution companies are failing to meet up with their role in providing free meters to Nigerians as stated in the Meter Asset Provider services (MAPs) of 2018, but have engaged in corrupt collection of money from customers for prepaid meters without installing the meters. Also, x-raying the prepaid meters challenge in Nigeria, Olalere [54] stated that the DISCOs themselves are a major problem in the optimization of the benefits of pre-paid metering. The author opined that DISCOs are exhibiting lethargy in implementing metering measures by "gaming" the system, which they blame on massive unpaid bills at the time of takeover, inadequate data on the Nigeria electricity supply industry (NESI), significant electricity theft, bypassing of meters by consumers, high inflation rate, unstable exchange rate, generation & transmission low capacities, high gas price and lack of orientation of the consumers.

Moreover, in the reports of Adekoya [5] and James–Igbinadolor [41], many customers blames the DISCOs for what they called "unholy alliances with estimated billing" for deliberately not supplying prepaid meters to customers in order to slap customers with outrageous bills that have no bearing whatsoever on the quality or quantity of electricity supplied to customer. The non-metering issue by the DISCOs is a major source of revenue loss in Nigeria according to the Nigerian extractive industry transparency initiatives, NEITI [48]. There is therefore a need for optimizing market efficiency as seen in the telecommunication industry [41].

Furthermore, one of the challenge of the prepaid metering system in Nigeria is the expansive power granted to the Distribution companies in the 2018 Meter Asset Providers (MAP) regulations [27], which led to wide discretionary powers given to DISCOs to declare a metering gap, thereby creating room for corruption and a situation whereby private entrants can only participate in metering when it does not have conflicting interests with DISCOs. Also, another problem with the wide power granted to DISCOS is that consumers whilst paying for metering service charge and cost of meters are at the mercy of MAP and DISCOs. Others challenges according to the author includes; lack of clarity with respect to ownership of meter after the expiration of the technical life of the meters; lack of knowledge about consumers obligations under the MAP arrangement and also the MAP regulation makes no case for consumers who paid for meters under the CAPMI scheme but were not provided.

5.2.5 Future Prognosis

In a bid to solve the many metering challenges in Nigeria, Tsado et al. [65] advocates for the use of an advance prepaid metering energy meter, using Global System for Mobile Communications (GSM) technology. According to the authors, this new device provides high level energy management through an advanced algorithm embedded in a micro-controller, monitoring peak and off– peak energy consumption. This new prepaid energy meter is designed to communicate to the consumer through a GSM module as a Short Message Service to the registered phone number programmed with the micro controller.

Supporting the usage of GSM based prepaid meters, Dike et al. [22] had stated that this type of prepaid meters will minimize household electricity theft in Nigeria, which entails meter tampering, meter bypass, illegal terminal taps of overhead lines on the low tension side of the transformer, illegal tapping to bare wires or underground cables, billing irregularities and unpaid bills.

However, Aniedu [11] stated that with the disadvantages of prepaid meters in Nigeria such as travelling to designated locations to buy the tokens (either the keypad type or smartcard type), smart meters are preferable. The authors argued that smart meters are like prepaid meters but with additional features such as ability to communicate with other meters, ability to monitor and control energy usage of home appliances and capability for remote monitoring and management (such as reconnection, disconnection and credit recharge). Moreover, smart meters utilizes technologies such as Bluetooth for Home Area Networks (HAN) employing the IEEE 802.15.1 protocol, Broadband power line communication (BPL) employing TCP/IP over radio frequency spectrum, WI-FI or WI.MAX technology employing the 802.11a/b/g/n standard, the Global system for mobile communication (GSM) and the General packet radio service (GPRS).

Furthermore, Akpan et al. [8] in supporting the mass metering system through the usage of Smart meters opined that smart meters with a prepayment functionality enables new and flexible methods of crediting the consumers account and it could be done remotely, using suitable payment means. It is a system that can disconnect when predetermined energy usage thresholds are reached.

However, irrespective of the prepaid metering choice, the Federal Government of Nigeria in October, 2020 launched the National Mass Metering Program (NMMP) implementation to bridge the metering gap in the country [18]. The NMMP enables the Central Bank of Nigeria (CBN) to financially support DISCOs and local meter manufacturers. NMMP has the following objectives [18]:

a.   Increase Nigeria's metering rate
b.   Elimination of arbitrary estimated billing
c.   Strengthen the local meter value chain by increasing local meter manufacturing, assembly and deployment capacity.
d.   Support Nigeria's economic recovery by creating jobs in the local meter value chain
e.   Reduction of collection losses and increasing financial flows to achieve 100% market remittance obligations of the DISCOs and
f.   Improve network monitoring capability and availability of data for market administration and investment decision making.

5.2.6   Highlights

- Prepaid meters were launched in Ghana and Nigeria (two West African nations) to foster financial probity and solve issues such as reducing collection expenses, overbilling of customers, efficient cash flows, eliminating bad debts, and improving revenue collection.
- Significant challenges exist during the prepaid journey in Ghana and Nigeria ranging from technical problems (voltage level drops, single-phase overloading, etc.), logistic problems (meter installation delays, lack of local manufacturing capacity, etc.) to financial problems (high cost of prepaid meters, etc.).
- There is the need to leverage further technological innovations like smart prepaid metering, big data analytics and machine learning algorithms in resolving some of the technical, logistical and financial challenges the roll-out of prepaid meters is experiencing in these West African nations.

References

1. Acakpovi A, Abubakar R, Asabere NY, Majeed IB (2019) Barriers and prospects of smart grid adoption in Ghana. In: 2nd international conference on sustainable materials processing and manufacturing (SMPM 2019). Procedia Manuf 35:1240–1249
2. Acakpovi A, Asabere NY (2018) Modern electrical grid optimization with the integration of big data and artificial intelligence techniques, 29
3. Adams AA (2019) Investigation: prepaid electricity meters—a scheme that became a scam. The Cable. Available at https://www.thecable.ng/investigation-prepaid-electricity-meters-a-scheme-that-became-a-scam. Accessed 15 Nov 2020
4. Adekitan AI, Adetokun BB, Aligbe A, Shomefun T, Orimogunje A (2018) Data based investigation of the energy metering type, billing and usage of sampled residents of Ota community in Nigeria. Data in Brief. https://doi.org/10.1016/j.dib.2018.07.047

5. Adekoya F (2019) New challenges in deploying meters to consumers. The Guardian. Available https://guardian.ng/energy/new-challenges-in-deploying-meters-to-consumers/. Accessed 15 Nov 2020

6. Ahmed A, Gong J (2017) Assessment of the electricity generation mix in Ghana. The potential of renewable energy. Degree project in technology, first cycle, 15 credits. Stockholm, Sweden. Kth Royal Institute of Technology, School of Architecture and the Built Environment. www.kth.se

7. Ajenikoko GA, Adelusi LO (2015) Impact of prepaid energy metering system on the electricity consumption in Ogbomoso South local government area of Oyo State. Comput Eng Intell Syst 6:99–106

8. Akpan AG, Shedrack M, Barida B (2019) Smart metering intelligent systems: a Panacea for efficient energy management in Nigeria. Restaur Bus 15:62–69

9. Alfred L (2011) Prepaid meter—a bane to the Ordinary Ghanaian. GhanaWeb. Available at https://www.ghanaweb.com/GhanaHomePage/features/Prepaid-meter-a-bane-to-the-ordinary-Ghanaian-209234. Accessed 25 Nov 2020

10. Amhenrior HE, Edeko FO, Ogujor EA, Emagbetere JO (2017) A comparative analysis of prepayment meters on selected loads. J Nigerian Assoc Math Phys 41:397–406

11. Aniedu AN, Inyiama HC, Azubogu ACO (2017) Smart meters and advanced metering infrastructure: Panacea to Nigeria's energy billing and monitoring problem. J Eng Appl Sci 1, JEAS ISSN:1119–8109

12. Armah AR (2016) Smart metering data integrity infrastructure for utility grid Management systems—a case study of base transceiver stations. Master Degree Dissertation. Kwame Nkrumah University of Science and Technology

13. Asante E (2014) Benefits and challenges of ECG prepaid metering. Graphic communications. Available at https://www.graphic.com.gh/features/opinion/benefits-and-challenges-of-ecg-prepaid-metering.html. Accessed: 15 Nov 2020

14. Azila–Gbettor E, Atatsi EA, Deynu F (2015) An exploratory study of effects of prepaid metering and energy related behavior among Ghanaian household. Int J Sustain Energy Environ Res 4(1):8–21. 10.18488/journal.13/2015.4.1/13.1.8.21

15. Boadu MTB (2015) Assessing customer satisfaction of Pre-payment meter usage in Asokwa district of ECG in Kumasi Metropolis. Master Degree Thesis. Kwame Nkrumah University of Science and Technology (KNUST). Ghana

16. Britannica (2020) Ghana. Available at https://www.britannica.com/place/Ghana. Accessed 18 Nov 2020

17. CBN (2015) Analysis of energy market conditions in Nigeria. Central Bank of Nigeria. Occasional paper no. 55

18. CBN (2020) Framework for financing of National Mass Metering Programme (NMMP). Central Bank of Nigeria. October 2020

19. CIA World Fact Book (2020) Central intelligence agency, World Factbook. Library Publications. Ghana. Available at https://www.cia.gov/library/publications/the-world-factbook/geos/gh.html. Accessed 12 Nov 2020

20. CIA World Fact Book (2020) Central intelligence agency, World Factbook. Library Publications. Nigeria. Available at https://www.cia.gov/library/publications/the-world-factbook/geos/ni.html. Accessed: 15 Nov 2020

21. Common Wealth (2020) Ghana key facts. Available at https://thecommonwealth.org/our-member-countries/ghana. Accessed 15 Nov 2020

22. Dike DO, Obiora UA, Nwokorie EC (2015) Minimizing household electricity theft in Nigeria using GSM based prepaid meter. Am J Eng Res (AJER) 4(1):59–69

23. ECG (2013) Electricity company of Ghana. Proposal for review in distribution service charge. Available at https://www.purc.com.gh/purc/sites/default/files/Tariff_proposal_for_2013_ECG.pdf. Accessed 15 Nov 2020

24. ECG website (2020) Electricity company of Ghana Limited. Available at https://www.ecggh.com/index.php/customers-service/meter-information/prepayment-metering-404. Accessed 15 Nov 2020

25. Ehun ME, Amoako–Tuffour J (2016) A review of the trends in Ghana's power sector. Energy Sustain Soc. https://energysustainsoc.biomedcentral.com/articles/10.1188/s13705-018-0075-y
26. Energy Commission (2020) Who we are. Available at http://www.energycom.gov.gh. Accessed 15 Nov 2020
27. Eze UG (2019) An appraisal of the legal implication of the Meter Asset Providers (MAP) regulation 2018 for the protection of electricity consumers in Nigeria. IRLJ 1(2):22–29
28. Fagbohun OO, Femi–Jemilohun OJ (2017) Prepaid metering empowerment for reliable billing and energy management of electricity consumers in Nigeria. J Sci Res Reports 17(2):1–13, JSRR.36344
29. FocusEconomics (2020). Nigeria economic outlook. Available at https://www.focus-economics.com/countries/nigeria. Accessed 15 Nov 2020
30. FocusEconomy (2020) Ghana economic outlook. Available at https://www.focus-economics.com/countries/ghana. Accessed 15 Nov 2020
31. GetInvest (2020a) Ghana energy sector. Available at https://www.get-invest.eu/market-information/ghana/energy-sector/. Accessed 15 Nov 2020
32. GetInvest (2020b) Nigeria energy sector. Available at https://www.get-invest-eu/market-information/nigeria/energy-sector/. Accessed 15 Nov 2020
33. Electricity Supply Plan Committee of Ghana, 2019 Electricity Supply Plan for the Ghana Power System. Available at http://www.energycom.gov.gh/files/2019%20Electricity%20Supply%20Plan.pdf (Accessed: 15 November 2020)
34. Ghana Energy Mix Report (2019) Ghana's ECG installs 4, 000 prepaid meters in Kibi. Available at https://www.energymixreport.com/ghanas-ecg-installs-4000-prepaid-meters-kibi/. Accessed 15 Nov 2020
35. Ministry of Energy, Ghana (2020) Overview of the Ghana power sector. Available at https://www.energymin.gov.gh/sector-overview. Accessed 15 Nov 2020
36. Global Legal Insights (2020) Energy 2021|Ghana GLI. Available at https://www.globallegalinsights.com/practice-area/energy-laws-and-regulations/ghana. Accessed 15 Nov 2020
37. GRINCO (2020) Functions. Ghana grid Company Limited. Available at https://www.gridcogh.com/home/overview/. Accessed 15 Nov 2020
38. Harvey E (2005) Managing the poor by remote control: Johannesburg's experiments with prepaid water meters. The age of commodity: water privatization in Southern Africa
39. IDE–JETRO (2020) Electricity Company of Ghana (ECG). Data & resources. Institute of Developing Economies Japan External Trade Organization. Available at https://www.ide.go.jp/English/Data/Africa_file/Company/ghana02.html. Accessed 15 Nov 2020
40. IEA, World Energy Outlook (2020) Electricity access. Available at https://www.iea.org/reports/sdg7-data-and-projections/access-to-electricity. Accessed 15 Nov 2020
41. James–Igbinador N (2020) Discos' unholy affinity with estimated billing. ThisDayLive. Available at https://www.thisdaylive.com/index.php/2020/06/14/discos-unholy-affinity-with-estimated-billing/ Accessed 15 Nov 2020
42. Kettless PM (2004) Prepayment metering systems for the low income group. PRI Ltd, London
43. Kimathi Partners (2018) Electricity regulation in Ghana. Kimathi & Partners Corporate Attorneys. Available at https://www.lexology.com/library/detail.aspx?g=910beca2-f3bf-420f-8597-e7e3e6f53b38. Accessed 15 Nov 2020
44. Kwofi M (2019) Ghana electrometer supports communities with 790, 000 meters. Graphic. Available at https://www.graphic.com.gh/business/business-news/ghana-electrometer-supportss-communities-with-790-000-meters.html. Accessed 15 Nov 2020
45. Makanjuola NT, Shoewu O, Akinyemi LA, Ajose Y (2015) Investigating the problems of prepaid metering systems in Nigeria. Pacif J Sci Technol 16(2, November Fall):22–31
46. Malama A, Mudenda P, Ng'ombe A, Makashini L, Abanda H (2014) The effects of the introduction of prepayment meters on the energy usage behavior of different housing consumer groups in Kitwe, Zambia. AIMS Energy 2(3):237–259. 10.3934/energy.2014.3.237
47. NERC (2020) Electricity on demand. First Quarter, 2020: Nigerian electricity regulatory Commission. Available at https://nerc.gov.ng/index.php/library/documents/NERC-Reports/NERC-Quarterly-Reports/NERC-First-Quarter-Report-2020/. Accessed 15 Nov 2020

48. Nextier (2018) Addressing the challenges of electricity metering in Nigeria. Nextier Power. Available at https://www.nigeriaelectricityhub.com/2018/02/02/addressing-the-challenges-of-electricity-metering-in-nigeria/. Accessed 15 Nov 2020
49. Nordea Market (2020) The economic context of Nigeria. Economic and Political overview, Nigeria. Available at https://www.nordeatrade.com/en/explore-new-market/nigeria/economical-context. Accessed 15 Nov 2020
50. Nwani SE, Ozegbe EA, Olunlade YT (2018) An overview of Nigerian energy sector: prospects and challenges. Available at https://www.academia.edu/37834838/AN_OVERVIEW_OF_NIGERIAN_ENERGY_SECTOR_PROSPECTS_AND_CHALLENGES_BY. Accessed 15 Nov 2020
51. O'Sullivan KC, Viggers HE, Howden–Chapman PL (2014) The influence of electricity prepayment meters use on household energy behavior. Sustain Cities Soc. https://dx.doi.org/10.1016/j.scs.2013.10.004
52. Ofonyelu CC, Eguabor RE (2014) Metered and unmetered billing: how asymmetric are the PHCN Bills? J Soc Econ Res 1(5):97–107
53. Ogbuefi UC, Ene PC, Okoro PA (2019) Prepaid meter tariffing for actual power consumption in an average household: a case study of Nigeria DISCOs. Nigerian J Technol (NIJOTECH) 38(3):750–755. https://doi.org/10.4314/njt.v38i3.29
54. Olalere PO (2018) Are the Discos complying with their performance agreement on metering the Nigerian electricity consumers? SPA Ajibade & Co. Energy & Natural Resources
55. Oluleye FA, Koginam AO (2019) Nigeria's energy sector privatization: reforms, challenges and prospects. South Asian Res J Hum Soc Sci 1(2):189–197. https://doi.org/10.36346/SARJHSS.2019.v01i02.028
56. Orient Energy Review (2020) Electricity consumers to pay more for prepaid meters. Available at https://www.orientenergyreview.com/power-sector/16356/. Accessed 15 Nov 2020
57. Orukpe PE, Agbontaen FO (2013) Prepaid meter in Nigeria: the story so far and the way forward. Adv Mater Res 824(2013):114–119
58. Our World in Data (2020) Per capital electricity consumption. Available at https://ourworldindata.org/energy/country/ghana?country=GHA. Accessed 15 Nov 2020
59. Pombo-van Zyl N (2014) Ghana introduces prepaid smart meters. ESI Africa. Available at https://www.esi-africa.com/industry-sectors/energy-efficiency/ghana-introduces-prepaid-smart-meters/. Accessed 15 Nov 2020
60. PURC (2020) Who we are. Public Utilities Regulatory Commission (PURC). Available at http://www.purc.com.gh/purc/features/fusion-menu. Accessed 15 Nov 2020
61. PWC (2016) Powering Nigeria for the future. The Power Sector in Nigeria. Pwc publications
62. Quayson–Dadzie J (2011) Customer perception and acceptability on the use of prepaid metering system in Accra West Region of Electricity Company of Ghana. Commonwealth Executive Masters Thesis. Institute of Distance Learning, Kwame Nkrumah University of Science and Technology, Ghana
63. Simpson JC (1996) 'Paying as we go—ahead'. In Proceedings of a colloquium on UK electricity prepayments systems. Institution of Electrical Engineering's Digest No. 1996/047, pp 2/1–2/5
64. TradingEconomics (2020) Nigeria—Electric Power Consumption (Kwh per Capita). Available at https://tradingeconomics.com/nigeria/electric-power-consumption-kwh-per-capita-wb-data.html. Accessed 15 Nov 2020
65. Tsado J, Mustapha BM, Ndakara AI (2017) Enhancement of electrical energy transaction through the development of a prepaid energy meter using GSM technology. Int J Res Studies Electr Electron Eng(IJRSEEE) 3(4):10–18. http://dx.doi.org/10.20431/2454-9436.0304003
66. Tuffour M, Sedegah DD, Asante K, Bonsu D (2018) The role of Pre-paid meters in energy efficiency promotion: merits and demerits in Accra, Ghana. Int J Eng Trends Technol (IJETT) 60(1):57–64
67. Ubogu S (2015) Country report on Nigeria energy system. IEEJ, Ministry of Energy, Asaba, Delta State, Nigeria
68. Ufondu I, Ibeku II, Vidal S (2020) Electricity regulation in Nigeria. Lexology. Available at https://www.lexology.com/library/detail.aspx?=88f16847-8f84-46e4-9b33-92b4f477e23f. Accessed 15 Nov 2020

69. UK Power Limited (2012) A simple guide to gas and electricity meters. http://www.ukpower.co.uk/home_energy/guide_to_gas_electricity_meters. Accessed 15 Nov 2020

70. VanguardNews (2019) Nigeria needs 24.7 m prepaid meters—expert. Available at https://www.vanguardngr.com/2019/08/nigeria-needs-24-7m-prepaid-meters-expert/. Accessed 15 Nov 2020

71. VRA (2020) Profile of VRA. Volta River authority. Available at https://www.vra.com/about_us/profile.php

72. World Bank (2020a) Population total—Ghana. The World Bank. Available at https://data.worldbank.org/indicator/SP.POP.TOTL?locations=GH. Accessed 15 Nov 2020

73. World Bank (2020b) Population, total—Nigeria. Available at https://data.worldbank.org/indicator/SP.POP.TOTL?locations=NG. Accessed 15 Nov 2020

74. World Population Review (2020) Population of cities in Ghana. World Population Rev. Available at https://worldpopulationreview.com/countries/cities/ghana. Accessed 15 Nov 2020

75. Worldometer (2020a) Ghana population. Available at https://www.worldometers.info/world-population/ghana-population. Accessed 15 Nov 2020

76. Worldometer (2020b) Nigeria population. Available at https://www.wordometers.info/world-population/nigeria-population/. Accessed 15 Nov 2020

77. Zaglago L, Jimoh B, Leo RD (2019) Drivers of smart grid technology in Ghana. In: Proceedings of the world congress on engineering and computer science, 2019. WCECS 2019, October 22–24, 2019 San Francisco, USA

Chapter 6
Eastern Africa Region

6.1 Tanzania

6.1.1 Geographical Overview

The United Republic of Tanzania is a country located in East Africa bordering the Indian Ocean between Kenya and Mozambique, with geographical coordinates of 6°00S and 35°00E [10]. Tanzania's total land area is 947, 300 km^2 comprising of 885, 800 km^2 land area and 61, 500 km^2 water, with the following neighboring countries; Burundi, Kenya, Malawi, Mozambique, Democratic Republic of Congo, Rwanda, Uganda and Zambia as shown in Fig. 6.1. Tanzania is the sixth most populous country in Sub—Saharan Africa, connecting six land—locked countries to the Indian ocean and is home to Africa's highest mountain of Kilimanjaro with several wild life national parks [3, 37].

According to Worldometer [51] based on extracted data from the United Nations, Tanzania population as of November 2020 is 60 465 128 with a population density of 67 per km^2 and a life expectancy of 66.39 years of which the city with the highest population in Tanzania is Dar es Salaam, estimated at 2 698 652. The Nation's capital is Dodoma and its official languages are Swahili, English Language and Arabic, with religious beliefs such as Christianity, Islam and traditional faith [9].

Economically, Tanzania is a middle income growing country with a Gross National Income (GNI) per capita of $1,080 as of 2019 [49]. According to Statista [40] and Macrotrends [25], Tanzania's GDP is 198.65 USD as of 2020 and a GDP per capita of $1,122 as of 2019. The CIA World Factbook [9] stated that the main industries in the Tanzanian economy are agricultural processing (Sugar, Beer, Cigarettes, Sisal twine); mining (diamonds, iron & gold), salt, soda ash, oil refining, shoes, apparel, fertilizers and cement.

© The Author(s), under exclusive license to Springer Nature Switzerland AG 2021   103
N. Kambule and N. Nwulu, *The Deployment of Prepaid Electricity Meters in Sub-Saharan Africa*, Lecture Notes in Electrical Engineering 759,
https://doi.org/10.1007/978-3-030-71217-4_6

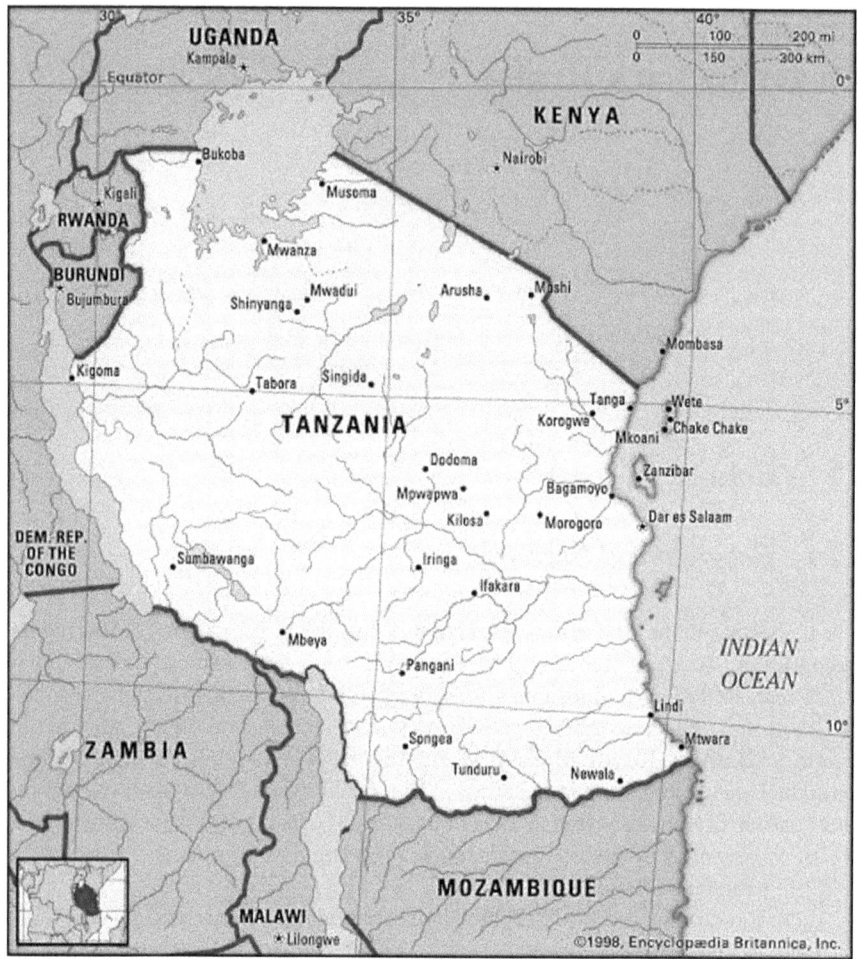

Fig. 6.1 The United Republic of Tanzania [5]

6.1.2   Energy Overview

Energy is an indispensable input that ensures sustainable development in a country [19]. Tanzania's energy sources include wood fuel, hydropower, biomass energy, natural gas, coal, uranium, and solar energy [13]. The country possesses very large reserves of black coal (1.9 billion tons) and gas (1.6 billion m^3) with large sources of renewable energy [13, 17, 36]. Tanzania's electrical energy is derived mainly from hydroelectric sources and natural gas resources [17]. According to Netherland Energy Agency (NEA) [36], electrical energy is supplied from oil (1381 GWh), Gas (2764 GWh), Biofuels (21 GWh), Hydro (2108 GWh), solar PV (21 GWh) totaling 6295 GWh.

Table 6.1 Generating plants in Tanzania and their owners [17]

Power station	Owner	Installed capacity (MW)
Hydro power plants		
Kidatu	TANESCO	204
Kihansi	TANESCO	180
Mtera	TANESCO	80
New Pangani Falls	TANESCO	68
Hale	TANESCO	21
Nyumb Ya Mungu	TANESCO	8
Uwemba	TANESCO	0.84
Mwenga	Mwenga Hydro Ltd	4
Yovi	SPP	0.95
Gas power plants		
Songas	IPP	189
Ubungo 1 gas plants	TANESCO	102
Tegeta gas plants	TANESCO	45
Ubungo 11 gas plants	TANESCO	121
Kinyerezi 1	TANESCO	150
Liquid fuel power plants		
IPTL	IPP	103
Diesel (TANESCO)	TANESCO	7.4
Nyakato	TANESCO	63
Biomass power plants		
TANWAT	1PP	1.5
TPC	IPP	9
Total		1357.69

Table 6.2 Kenya electricity's consumption [35]

Final electricity consumption	GWh
Industry	4229
Residential	2544
Commercial and Public service	1153
Total	7926

According to Power Africa [37], installed electricity capacity is 1504 MW, comprising of Hydroelectric (568 MW), Thermal (925 MW) and other renewables 82.4 MW. The generated electricity is managed by the Tanzania Electric Supply Company Limited (TANESCO), which is in charge of the generation, transmission and distribution of electricity [36]. Also, GetInvest [17] citing the Tanzania Ministry of Energy and Mineral 2016 Quarterly Digest stated that most electricity generating plants are owned by TANESCO, Small Power Producers (SPP) and Independent Power Producers (IPP) as shown in Table 6.1.

Furthermore, for the transmission and distribution of the generated electricity by TANESCO, there are 48 transmission substations interconnected by 7 transmission lines consisting of 3340 km of 220 kV, 2063 km of 132 kV, 668 km of 66 kV, 24 165 km of 33 kV, 6006 km of 11 kV and 71 629 km of 400 kV and 230 kV lines [35]. Also, TANESCO generates, transmits, distributes, purchase and sells electricity to the Tanzania mainland and also sells bulk power to the Zanzibar Electricity Corporation (ZECO), which also sells to the public [41].

According to GetInvest [17], the economic sectors that consume electricity as of 2014 are industry (25.5%, 1 270 GWh), Residential (44.8%, 2227 GWh), commercial and public services (22.9%, 1141 GWh), Agriculture/Forestry (3.6%, 180 GWh), others non-specified (3.1%, 158 GWh), totaling 4,976 GWh. with an electricity per capita consumption of 137 kWh/year. Furthermore, electricity access in Tanzania is at 40%. Out of this number, 71% are in urban areas and 23% in rural areas, whilst 35 Million of the populace is without access to electricity [18].

The Tanzanian energy sector is managed by the following stakeholders [4, 36]:

a. Ministry of Energy and Minerals (MEM): this is the Tanzanian Ministry saddled with the responsibility of coordinating activities involving energy in the country through the creation of policies, frameworks, strategies and guidelines for optimal utilization of the sectors potentials, especially renewable energy.

b. Rural Energy Agency (REA): this is an autonomous agency that is responsible for the implementation of electrification projects in rural areas of Tanzania through the financial utilization of Rural Energy Fund. The agency is coordinated by the Rural Energy board, consisting of eight delegates from different government agencies and civil societies.

c. Energy and water utilities regulating authority: this is Tanzania's energy regulator, which is autonomous in nature and a multisectoral regulating authority. The agency is responsible for regulation of the Tanzania energy sector and regulation of electricity tariffs.

d. TANESCO: this is a government owned utility, which is responsible for generation, transmission and distribution of electricity to the consumer. Also, the utility buys electricity from different generating companies such as Independent Power Tanzania limited (IPTL), Symbion Company and Agrreko Company.

e. Other power producers: in Tanzania, there are three different groups of power producers, which contribute 40% to the installed capacity, therefore helping TANESCO in power supply. The three groups are Independent Power Producers (IPP), Emergency Power Producers (EPP) and Small Power Producers (SPP).

The IPPs runs 9 large power plants in Tanzania usually larger than 10 MW. There are also two EPP namely; Aggreko and Symbion LLC. SPP are private companies that run power plants less than 10 MW.

f.   Zanzibar Electricity Corporation (ZECO): this is a government owned utility that facilitates the generation, transmission, transformation, distribution, supply and use of electricity in the Island of Zanzibar, through a working relationship with TANESCO.

Furthermore, laws, regulations and policies that guides and regulates the Tanzania energy sectors are:

* National Energy Policy 2015
* The Model Power Purchase Agreement
* Public Private Partnership Policy 2009
* Public—Private Partnership Act, 2010
* Energy and Water Utilities Authority Act 2001 and 2006
* Rural Energy Act 2005
* Electricity Act 2008
* The Petroleum Act 2008
* Standardized Power Purchase Agreement and Tarriffs (2008)
* The Electricity Supply Reform Strategy and Road Map 2014–2015
* The Electricity (Market Re-Organization and Promotion of Competition) Regulations 2016
* Power Systems Master Plan (PSMP) 2012
* MEM Strategic Plan from 2011/12–2015/16
* Scaling up Renewable Energy Programs (SREP)—Investment Plan for Tanzania (May 2013)
* Biomass Energy Strategy (BEST) for Tanzania
* Electricity Supply Industry (ESI) Reform Strategy and Roadmap, 2014–2025.

6.1.3   Prepaid Electricity History and Current Status

Mushi [32] states that prepaid meters   are electricity meters where consumers pay before using electricity. Also, Magambo [26] opined that prepaid meters are meters that automatically disconnect a consumer, when the purchased electricity credit has finished. Mhando [29] stated that prepaid metering was introduced in Tanzania by TANESCO for operational efficiency and revenue collection effectiveness. The factors that necessitated the deployment of prepaid meters are as follows; delayed generation of bills for credit type meters, non-payment of monthly bills, energy theft by customers and high cost of disconnection campaigns for debt collection.

In Tanzania, the prepaid metering system was introduced by TANESCO in a pilot project from 1995 to 2004 in four regions within the city of Dar es Salaam, Tanzania's highest populated city [29, 24]. The prepaid system when introduced was targeted at small power users either on single or three phase current meters consuming less than 7600 kWh and was successful due to an improvement in revenue collection and significant reduction in operational cost. However, Magambo [26] and Sospeter [39] opined that the prepayment metering project started between 1993 and 1997 through a world bank funded program called "Lipa Umeme Kadiri Utumiavyo" (LUKU), translated in English as "Pay for electricity as you need it", which led to 40,622 prepaid meters installed. The success of the pilot project led to the roll out of prepaid metering system throughout the country, which ensured that customers controlled their energy consumption by 8%, using energy as necessary.

Mushi [32], Magambo [26] and Sospeter [39] in x-raying the importance and benefits of the prepaid meters grouped them into benefits to the TANESCO and to the consumers. The benefits of prepaid meters to TANESCO are:

- Possibility of tracking individual and group sales.
- Quick and early detection of fraud related activity.
- Provision of comprehensive management reports.
- Up-front payments.
- Elimination of bad debts.
- Elimination of disconnection and reconnection fee.
- Improved customer services.
- There are no more incorrect and unpaid bills.
- Efficient and effective load control.

Also, the benefits and importance of prepaid metering to the customers are:

- Controlled electricity usage.
- Effective budget management.
- Convenience in buying the electricity.
- No cost and anxiety for reconnection and disconnection.
- Improved energy consciousness.

Mushi [32] further stated that the process of prepaid metering in Tanzania includes the following:

- Consumer's buys electricity from TANESCO vending stations, telephone company services like Vodacom and from ATM machines.
- The consumer receives a twenty (20) digit codes, which is to be keyed into the meter.
- The codes are inputted into the meter.
- The meter is credited and is gradually depleted depending on usage and speed of consumption.

However, despite the benefits of prepaid metering system in Tanzania, as of 2009, there are 520 000 credit meters which needs to be replaced with prepaid meters, leading to a yearly target of 100 000 replacements of credit meters with prepaid meters [29]. TANESCO in aligning and recognizing the benefits of prepaid metering system opined that most of its customers should use prepaid plans [27]. However, with the call for bids for prepaid electricity meter procurement for the 2018/2019 financial year by Zanzibar Electricity corporation (ZECO), it shows that the 100 000 yearly target of 2009 has not been met and about half of electricity consumers in Tanzania do not have prepaid meters [7].

6.1.4 Challenges of Prepaid Metering

The prepaid metering system in Tanzania despite its benefits has some deficiencies especially as it could not stop energy theft in the system [29]. Mogambo [26] and Mushi [32] stated that the problems faced with prepaid meters in Tanzania are:

- Ignorance and lack of education amongst prepaid meter consumers due to the complexities of prepaid meters such as problems whilst inserting disposable magnetic tokens, (which damage the magnetic tokens) and also the issue of inputting the twenty digits tokens into the prepaid meters.
- There are defective meters that do not give electricity to consumers and also failure of vending machines.
- The ease of manipulating vending machines and the instability of the machines, which leads to fraud, thus reducing utility revenue targets.

Also, Sospeter [39] stated that sixteen (16) years after the introduction of prepaid meters in Tanzania, there still exists a huge loss by TANESCO through energy thefts and debts owed the utility, which prepaid meters could not solve due to meter bypassing, thefts and illegal connection.

6.1.5 Future Prognosis

The massive and rampant electricity theft in Tanzania despite the usage of prepaid meters has led to the adoption and usage of Automatic Meter Reading (AMR) systems, a subsystem of smart metering. This system in Tanzania was launched by TANESCO in 2007, through a pilot project, involving 100 large power users, which resulted in 62% improved sales and 73% revenue protection [29]. Mhando [29] and Mnzava [30] highlighted the benefits and advantages of the AMR system:

a. Non-paying customers can be remotely disconnected.
b. There is an alarm, whenever there is a meter tampering attempt.
c. Improved quality of services, as temporary breakdown can be detected by the AMR, communicated and quickly fixed by the utility.
d. Low cost of billing due to automated billing process.
e. There is intelligent communications between the meters, customers and the TANESCO offices.

6.2 Kenya

6.2.1 Geographical Overview

According to the CIA World Fact Book [10] Kenya is a country located in Eastern Africa, whose capital is Nairobi with a land area of 569, 140 km^2 and water area of 11, 227 km^2, sharing borders with Ethiopia, Somalia, South Sudan, Tanzania and Uganda as shown in Fig. 6.2. The population of Kenya as of 2019, according to the World bank [49] was 52 573 973, however, according to Worldometer [51] based on United Nations projections, Kenya's Population as at November, 2020 is 54 239 315 with different ethnic groups and official languages like English Language and Kiswahili.

Economically, Kenya is an Agro-driven economy and it is the economic, financial and transport hub of East Africa, with economic growth averaging at 5.7% whilst it's the 9th largest economy in Africa [9, 50]. The economy majorly backed by Agriculture, Livestock, and Pastoral activities has a GDP of $95.3 Billion [15]. According to the Netherland Enterprise Agency (NEA) [35], Agriculture in Kenya involves the cultivation of tea, coffea, corn, wheat, sugarcane, fruit, vegetables, dairy

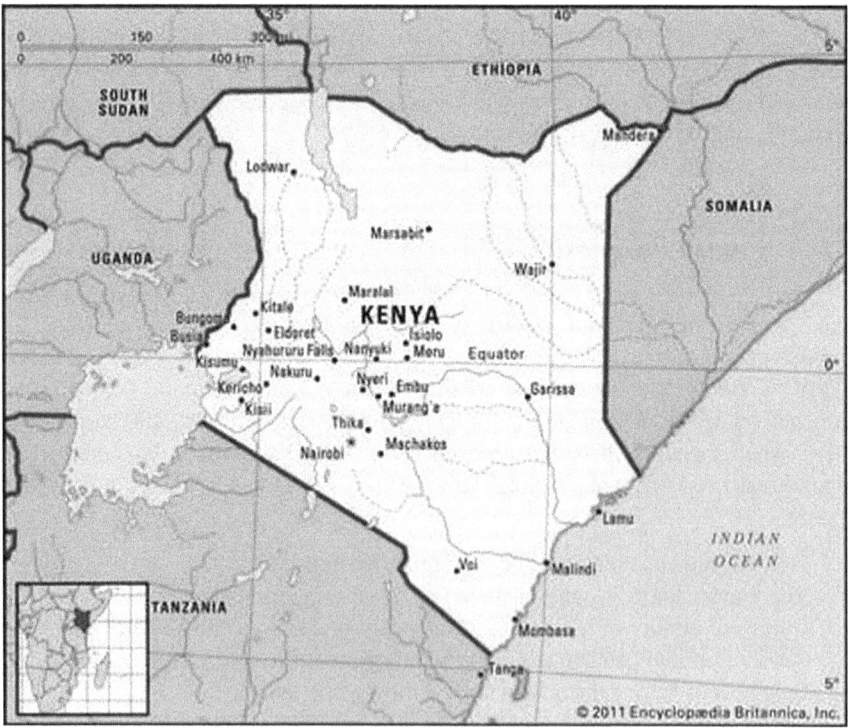

Fig. 6.2 The Republic of Kenya [6]

products, beef, fish, pork, poultry, eggs, while the industrial sector of the country comprises of plastic, textiles, battery, furniture, flour, clothing, aluminium, steel, lead, cement, commercial ship repair and tourism.

6.2.2 Energy Overview

Tellez and Waldron [43] opined that energy is central to human existence and it is one of the sustainable development goals (goal seven), which seeks to ensure access to affordable, reliable and sustainable modern energy for all. The economic growth and the rising national population of Kenya has increased public demand and exacerbated pressure on electricity supply, with an annual demand increase of 18.9% [36]. According to Energypedia [13], Kenya's energy sector comprises of petroleum and electricity, where wood fuel services the basic energy needs of most Kenyans. Since this study focuses on electricity, Kenya's electric energy is derived mainly from Hydropower, Fossil Fuels, Geothermal, Bagasse cogeneration, wind and solar energy [14, 36]. Also, according to International Energy Agency [18] and KPMG Sub-Saharan Africa power outlook [20], as of 2015, the proportion of electricity sources in Kenya are; Hydro (3310 GWh; 36%), Oil (1714 GWh, 19%), Geothermal (4,479 GWh, 44%), Biofuel (136 GWh, 1%), Wind (38 GWh, < 1%) and solar (1 GWh, <1%). Power Africa [36] in its reports opined that in comparing Kenya's population with its per—capita GDP, Kenya is performing well. Kenya has a per—capita power consumption of 161 kWh (as of 2014) compared to Nigeria which had a per-capita power consumption of 126 kWh, even though Nigeria's per—capita GDP is thrice that of Kenya. However, the per capita electricity consumption in Kenya as of 2019 is 222 kWh [2].

The Africa Energy Series [1] special report stated that electric power demand in Kenya is around 1802 MW as of 2018, 1 912 MW in 2019 and as of 2020 it ranges between 2600 and 3600 MW, rising at a rate of 3.6% annually. Moreover, according to the IEA [18] and NEA [35], electricity usage per sector is shown in Table 6.2.

Kenya's installed electricity capacity is approximately 2.7 GW, mainly from renewables such as Geothermal and Hydropower, of which geothermal sources have overtaken hydropower as Kenya's main source of electric power, thus ensuring energy availability during drought season [1]. The report further stated that Kenya Electricity Generating Company (Kenya Power), generates approximately 70% of installed capacity, while the remaining 30% is generated by Independent Power Producers (IPP), from the following firms; Westomont, AEP Energy Africa (Iberafrica), OrPower4Kenya Limited, Tsavo Power company, Aggreko and Africa's Geothermal international, operating in total 15 power plants. The generated electric energy is transmitted to different households using 4149 km of transmission lines of 200 kV or 132 kV. Currently approximately 4,500 km of new lines are currently being built with the introduction of 400 kV and 500 kV DC lines and three major regional interconnections to Ethiopia, Uganda and Tanzania in accordance with the Eastern African Power Pool [1, 20]. According to TradingEconomics [46] and IEA [18] world bank data, access to electricity is 75% as of 2018 and 84.5% as of 2019, where 99% are in the urban areas and 73% are in the rural areas, with 8 Million people without access to electricity.

In the electricity energy sub-sector in Kenya, as cited by NEA [35], Power Africa [20], CIA World Factbook [10] and Kenya National Electrification strategy [21], below are the summaries of relevant stakeholders:

- Ministry of Energy: the ministry is saddled with the stewardship of creating rules, regulations, and energy policies for the operational growth and investment optimization of energy in the country.
- Energy Regulatory Commission (ERC): This is an agency established by Kenya's Energy Act 2006, responsible for economic & technical leadership in conjunction with the Ministry of Energy and the regulation of the electric power, renewable energy and downstream petroleum sub-sector. The agency is responsible for enforcing licensing, settlement of disputes, setting and review of tariff, approving power purchase and network service contracts. The Agency is operationally independent and hence plays a key role in overseeing pricing and negotiation of power purchase agreements (PPAs) between Kenya Power and Power Producers.
- Rural Electrification Authority (REA): it's an agency established in 2007 that oversees the planning and commissioning of power plants for electrification of rural settlements in Kenya using the Kenya Rural Electrification Master Plan.
- Kenya Power and Lighting Company (KPLC) Ltd: this is Kenya's electric power brainbox, also known as Kenya power, established in 1975. The company is the middleman between Kenya electricity generator, transmitter and distributor to consumers. The company sells electricity to over 7.5 Million Kenyans as of June, 2020. The company is listed on the Nairobi securities exchange, with government having 50.1% controlling stakeholder's stake and private investors having 49.9% [23].
- Kenya Electricity Generating Company (KenGen): this company oversees the production and operation of power plants from hydro, geothermal, gas and diesel, accounting for 72% of electricity consumption in Kenya. It is owned by the government and private investors in the 70–30% proportion.
- Kenya Electricity Transmission Company (KETRACO): It's a government owned entity, responsible for planning, designing, developing, operating and maintaining all new transmission lines above 132 kV.
- Geothermal Development Company, GDC: The major Source of electricity in Kenya is Geothermal energy, hence a fully owned government company (GDC) established in 2008 is responsible for the exploration and production drilling of geothermal fields and also for the development of Steam fields.
- Independent Power Producers: these are private entities competing with KenGen for the production of electricity in Kenya, and there are 13 independent power producers in Kenya.

In the operation of these relevant stakeholders, there are laws and regulatory frameworks backing their operations in providing adequate power supply in Kenya, they include:

a. Energy Act 2006
b. Least Cost Power Development Plan (LCPDP) (2010)
c. Standardized Power Purchase Agreement
d. Electricity Licensing Regulations
e. Connection Guidelines for Small—Scale Renewable Generating Plants
f. Energy Management Regulation (2012).

6.2.3 Prepaid Electricity History and Current Status

Prepaid metering is an electricity billing system, where consumers pay in advance before consumption, and includes a superior electronic customer accounts management system. It is the opposite of post-paid metering [28]. Prepaid metering gives the customer control over their consumption rate and it is the upfront payment for electric usage [47]. To boost efficiency and improve quality of customer services, the prepaid meter was adopted and seeks to bring relief to the many challenges associated with the post-paid system. The challenges includes; lack of consumer control of consumption, ineffective revenue collection, inefficient coordination, high reconnection fees, overestimated bills and wrong estimation of bills, corruption during disconnection, and not returning consumers account deposits [18].

The prepaid metering system of electricity allows consumers to consume energy only when they have bought electricity units from Kenya Power and the supply is terminated when the credit units are exhausted. Consumers can be reconnected when those units are funded again. The prepaid meter helps Kenya Power in increasing managerial efficiency for distributing power to consumer, increases cash flow, reduces credit and arrears collection cost and avoiding reconnection charges [45]. Tirop [45] stated that the prepaid meter in Kenya works in the following steps; issuance of card with a unique meter number that must be presented when buying electricity at a sales outlet or mobile money platform, a credit slip or code, called token is received upon payment, the credit unit or code, which is usually 20—digits are loaded and the consumer receives electricity units for consumption.

This innovative metering system was adopted in Kenya and started on a pilot test through a contract with Actaris Ltd in April 2009 in Nairobi, Kisumu and Nakuru towns, while the full roll out was done in March, 2011 and within four (4) months 123, 000 prepaid meters had been installed throughout the country, mostly in Kenya's Capital city of Nairobi [22, 48]. Moreover, according to Datascience [12], Kenya power stated that as of March 20, 2017, there are 5.9 million customers connected to the National grid, of which 3.5 million are on prepaid meters, representing 59.32% on prepaid meters. Also, in April 2018, Kenya Power stated that it has 4.5 million customers on prepaid metering system. However, according to Bungane [8], Kenya Power in introducing its USSD code stated that its post-paid customers are 1.8 Million, showing a decrease in post-paid customers and an increase in prepaid customers.

In the usage of prepaid metering, customers pays for prepaid tokens using platforms such as mobile money transactions, KPLC offices, banks, sales Kiosks and Mobile and Web based self-service applications [44].

6.2.4 Challenges of Prepaid Metering

The introduction of prepaid metering system in Kenya in 2009 was intended to reduce and eliminate the shortcomings of the post-paid metering system. However, the installation and adoption of prepaid system led to the following challenges and shortcomings as noted by Wambua et al. [47] and Tirop [45]; faulty gadgets, poor consumer education on how to use the metering system, incorrect billings and high cost of connectivity. Moreover, according to the Nation [34] and Chege [11], whilst using the prepaid metering system, consumers in Kenya were dissatisfied due to the following challenges; limited places to buy tokens, delays in purchasing credits units through money transfer, long digits or codes and varying & inconsistent rates of units for the same amount (due to fuel price, foreign exchange & inflation).

Furthermore, another issue affecting the reliability of Kenya Power in terms of the prepaid metering system is electricity outages [42]. Another issue, is the shortage of pre-paid meters which is a major issue for both electricity consumers in Kenya and Kenya power; as customers wait for months to collect pre-paid electricity meters, they have paid for due to procurement issues by Kenya Power. The author reported a complaint of a customer in the following words "I have visited Kenya power offices over ten times and I have been informed that I have to wait longer as they seek the meters which will first be served to those who applied first" [31].

6.2.5 Future Prognosis

In tackling the bottlenecks and challenges surrounding the usage of prepaid meters, Kenya Power has opted for the use of smart meters [33]. This will improve transparency, accuracy and proper monitoring of electricity bills, outages and supplies. The Nairobi Garage [33] further reported that Kenya power has so far installed 15,736 smart meters for its large and small electricity consumers. According to Franek et al. [16], smart meters measures more variables related to electricity consumptions and also the quality of energy delivered. According to the authors, benefits of smart metering are the ability to control electric consumption rate, remote data acquisition, collection of data for analysis, reduction of illegal consumption and ability to schedule or on demand tariff charge and disconnect power supply.

Smart meters enable two—way communication between the meter and the control centre of the distribution company. The introduction of smart meters also enables the implementation of the presidential directive of President Uhuru Kenyatta on reduced electricity tariff for manufacturers who will be using power between 10 pm and 6 am

[33]. However, the main disadvantage of the smart meter is the higher cost of meters and need for communication infrastructure [16]. Furthermore, Kenya Power in 2014 earmarked $1.6 million for a two (2) year smart meter pilot to cut distribution and operating cost; thus saving the company $11 million over the next four (4) years [38].

6.3 Highlights

- The introduction of prepaid meters in Tanzania and Kenya stemmed from the need to give customers more control over their electricity use and the resultant bills. It also sought to eliminate incorrect or wrong meter readings.
- Even though prepaid meters have been installed for over 20 years ago in Tanzania and Kenya, challenges relating to meter reliability and meter bypassing/tampering remain and cause significant consumer dissatisfaction.
- Increased deployment of smart meters to combat the challenges in the prepaid electricity sector is crucial. There is also the need to increase the number of meters in the hands of those that desire them and work on the serious issues of access to electricity for significant portions of their populace.

References

1. Africa Energy Series (2020) Special report. Kenya. Invest in the Energy Sector of Kenya. https://www.africaoilandpower.com/africa-energy-series/kenya-special-report-2020/. Accessed 18 Nov 2020
2. Africa Power (2016) Development of Kenya's Power sector 2015–2020. USAID Kenya & East Africa, Nairobi, Kenya
3. BBC, 2020: Tanzania Country Profile. Available at https://www.google.com/amp/s/www.bbc.com/news/amp/world-africa-14095776 (Accessed on 18 November, 2020)
4. Bakary S (2018) The United Republic of Tanzania. Ministry of Energy, Country Report
5. Britannica (2020a) Kenya. https://www.britannica.com/place/Kenya. Accessed 12 Nov 2020
6. Britannica (2020b) Tanzania. https://www.britannica.com/place/Tanzania. Accessed 12 Nov 2020
7. Bungane B (2018) Zanzibar Electricity Corporation to procure electricity meters. ESI Africa. https://www.esi-africa.com/industry-sectors/metering/zanzibar-electricity-corporation-to-procure-electricity-meters/. Accessed 28 Nov 2020
8. Bungane B (2020) Kenya Power introduces USSD platform for customers to manage bills. ESI Africa. https://www.esi-africa.com/industry-sectors/finance-and-policy/kenya-power-introduces-ussd-platform-for-customers-to-manage-bills/. Accessed 18 Nov 2020
9. CIA World Fact Book (2020b) Central Intelligence Agency, World Factbook. Library Publications, Kenya. https://www.cia.gov/library/publications/the-world-factbook/geos/ke.html. Accessed 18 Nov 2020
10. CIA World Fact Book (2020a) Central Intelligence Agency, World Factbook. Library Publications, Tanzania. https://www.cia.gov/library/publications/the-world-factbook/geos/ke.html. Accessed 18 Nov 2020

11. Chege N (2012) Kenya Power's Pre-paid meters rated highly amid complaints. The Standard. https://www.google.com/amp/s/www.standardmedia.co.ke/amp/cartoon/article/200005 5303/kenya-powers-pre-paid-meters-rated-highly-amid-complaints. Accessed 23 Nov 2020
12. Datascience (2019) Approximately 6 Million Kenyan Households are connected to electricity. DataScience LTD. https://medium.com/@Datasciencing/approximately-6-million-kenyans-are-connected-to-electricity-62e55fc4d674. Accessed 18 Nov 2020
13. Energypedia (2020a) Kenya Energy Situation. https://energypedia.info/wiki/kenya_energy_situation. Accessed 18 Nov 2020
14. Energypedia (2020b) Tanzania Energy Situation. https://energypedia.info/wiki/Tanzania_Energy_situation. Accessed 18 Nov 2020
15. FocusEconomics (2020) Kenya Economic Outlook. https://www.google.com/url?sa=t&source=web&rct=j&url=https://www.focus-economics.com/countries/kenya&ved=2ahUKEwjSO8_vsortAhWEEFkFHSLFAGcQFjAOegQIBhAB&usg=AovVaw1X36gGnO47YAg8ZTkA2a-C. Accessed 22 Nov 2020
16. Franek L, Stasttny L, Fiedler S (2013) Prepaid energy in time of smart metering. In: 12th IFAC conference on programmable devices and embedded systems. The International federation of automatic control, pp 428–433
17. GetInvest (2020) Tanzania Energy Sector. Available at https://www.get-invest.eu/market-information/tanzania/energy-sector/ (Accessed on 18 November, 2020)
18. IEA (2020) International Energy Agency, Kenya Energy Outlook. Analysis from Africa Energy outlook 2019. https://www.iea.org/articles/kenya-energy-outlook. Accessed 18 Nov 2020
19. IRENA (2017) International Renewable Energy Agency. United Republic of Tanzania, Renewables Reading Assessment
20. KPMG Subsaharan Africa Power Outlook (2016) KPMG Africa Infrastructure. www.kpmg.com/africa. Accessed 18 Nov 2020
21. Kenya National Electrification strategy (2018) National Electrification Strategy: Key Highlights. 5 See KNES Investment Plan, Final Report, 2018. Accessed 18 Nov 2020
22. Kenya Power (2011) All you need to know about Kenya Power Prepaid Meters. https://kenyatech.wordpress.com/2011/10/20/all-you-need-to-know-about-kenya-power-prepaid-meters/. Accessed 18 Nov 2020
23. Kenya Power (2020) Kenya Power, Who we are. https://www.kplc.co.ke/content/item/14/about-kenya-power. Accessed 18 Nov 2020
24. Kumar Das D, Stern D (2020) Prepaid metering and electricity consumption in developing countries. Energy Insight. Applied Research Programme on Earth and Economic Growth, Energy and Economic Growth
25. Macrotrends (2020) Tanzania GDP Per Capita 1988–2020. https://www.macrotrends.net/countries/TZA/tanzania/gdp-per-capita. Accessed 18 Nov 2020
26. Magambo WDS (2000) Prepaid Metering in Tanzania. Smart Energy International. https://www.google.com/url?sa=t&source=web&rct=j&url=https://www.metering.com/wp-content/uploads/william%2520Magambo.pdf&ved=2ahUKEwj3gvzWoMXtAhWFUhUIHWhTBVsQFjAAegQIARAB&usg=AOvVaw38bJKjlXdPvN7ddyWKU7Pt. Accessed 25 Nov 2020
27. Makoye K (2017) Mobile Power Payments and Smart meters plug in Tanzanian homes. Reuters. https://www.google.com/amp/s/in.mobile.reuters.com/article/amp/idUSKCN1BI00B (Accessed on 24 November, 2020)
28. Mathenge P (2015) Influence of prepaid electricity meters adoption on the level of customer satisfaction: a case of Thika Sub County. Master Degree Research, University of Nairobi, Kenya, Kenya
29. Mhando WG (2009) Prepayment Expansions and AMR in Tanzania. Improving Operational Efficiency and Revenue Collection TANESCO presentation
30. Mnzava GEN (2012) Impact of AMR system on Revenue collection and Loss reduction. TANESCO Case Study, TANESCO Presentation
31. Murage G (2019) Kenya Power facing acute shortage of pre – paid meters. Star. https://www.google.com/amp/s/www.the-star.co.ke/amp/business/2019-10-02-kenya-power-facing-acute-shortage-of-pre-paid-meters/. Accessed 26 Nov 2020

32. Mushi PD (2014) Effectiveness of Pre-paid metering system in revenue collection: a case of TANESCO. Master of Science Dissertation. Mzumbe University Dar es Salaam, Tanzania
33. Nairobi Garage (2018) Smart Meters: New Digital Frontier for electricity distribution companies. https://nairobigarage.com/smart-meters-kenya/. Accessed 25 Nov 2020
34. Nation (2011) Prepaid electricity meters test the patience of consumers. Nation Africa. https://www.google.com/amp/s/nation.africa/kenya/life-and-style/money/prepaid-ele ctricity-meters-test-the-patience-of-consumers-787316%3fview=htmlamp. Accessed 18 Nov 2020
35. Netherland Energy Agency, NEA (2018a) Final Energy Report Kenya. Ministry of Foreign Affairs. Kees Mokveld & Steven Von Eije
36. Netherland Enterprise Agency, NEA (2018b) Final Energy Report Tanzania. Ministry of Foreign affairs. Kees Mokveld & Stephen von Eike. Version 6
37. Power Africa (2020) Tanzania Power Africa sheet. Tanzania Energy sector overview. US Agency for International Development. https://www.usaid.gov/powerafrica/tanzania. Accessed 18 Nov 2020
38. Smart Energy International (2014) Smart meters Africa: Kenya Power starts pilot for high—end consumers. Clarion Energy. https://www.smart-energy.com/regional-news/africa-middle-east/smart-meters-africa-kenya-power-starts-pilot-for-high-end-consumers/. Accessed 18 Nov 2020
39. Sospeter D (2020) Effectiveness of prepaid metering system in revenue collection: a case of TANESCO Meru District Arusha. Bachelor of Science Degree Research, University of Arusha
40. Statista (2020) GDP growth in Tanzania. Julia Farla. https://www.statista.com/statistics/113 3329/gdp-growth-in-tanzania-by-economic-activity/. Accessed 18 Nov 2020
41. TANESCO (2020) Tanzania Electric Supply Company Limited. Background. http://www.tan esco.co.tz/index.php/about-us/historical-background. Accessed 18 Nov 2020
42. Taneja J (2017) Measuring Electricity Reliability in Kenya. University of Massachusetts, STIMA Lab
43. Tellez C, Waldron D (2017) The fight for light. Improving Energy Access through Digital Payments. UN CDF, Better Than Cash Alliance. ISBN 978-1-946173-28-790000
44. Theron A (2018) Kenya power addresses prepaid and billing concerns. ESI Africa. https://www.esi-africa.com/regional-news/east-africa/kenya-power-addresses-prepaid-and-billing-concerns/. Accessed 18 Nov 2020
45. Tirop RK (2018) Prepaid electricity billing and the financial performance of Kenya power and Lighting Company. Kenyatta University, Master of Degree
46. TradingEconomics (2020) Kenya—Access to Electricity (% of Population). https://tradingec onomics.com/kenay/access-to-electricity-percent-of-population-wb-data.html. Accessed 18 Nov 2020
47. Wambua AM, Kihara P, Mwenemeru HK (2015) Adoption of prepaid electricity metering system and customer satisfaction. Int J Sci Res 6(9):1702–1710. https://doi.org/10.21275/ART 20176786
48. Wash L (2009) Actaris Awarded contract for Prepayment Pilot by KPLC. Itron Press Release. https://www.itron.com/it/company/newsroom/2016/06/10/actaris-awarded-contract-for-prepayment-pilot-by-kplc. Accessed 18 Nov 2020
49. World Bank (2020) The World Bank in Kenya. The World Bank Group. https://www.worldb ank.org/en/country/kenya/overview#1. Accessed 18 Nov 2020
50. World Bank (2020) The World Bank in Tanzania. Dar ES Salaam, Tanzania. https://www.wor ldbank.org/en/country/Tanzania/overview. Accessed 18 Nov 2020
51. Worldometer (2020) Tanzania Population. https://www.worldometers.info/world-population/tanzania-population/. Accessed 18 Nov 2020

Chapter 7
Southern African Region

7.1 South Africa

7.1.1 Geographical Overview

The Republic of South Africa is Africa's southernmost country with a great topography and cultural diversity. South Africa has a population of 59 602 202 (as of November, 2020), a total land area of 1 220 813 km^2 and a population density of 49 km^2, where 66.4% of her population resides in urban areas and 33.6% resides in rural areas [51]. According to the CIA World Fact book [7], South Africa shares boundaries with the following countries; Lesotho, Swaziland, Botswana, Namibia, Mozambique, and Zimbabwe as shown in Fig. 7.1. A majority black population (which constitutes 81% of the population) dominates the country whilst whites, coloured and the Indians/Asian population [25] constitute the rest.

The Republic of South Africa practices a constitutional multi-party democracy divided into three levels of governance: local, provincial and national governments, where the country's administrative headquarters is in Pretoria, the legislative headquarters is in Cape Town and the judicial headquarters is in Bloemfontein [25]. The Republic is a multicultural country with 11 official equal status languages, which are English, Tswana, Afrikaans, Tshivenda (Venda), Pedi, Sesotho (Sotho), IsiNdebele, Xitsonga (Tsonga), Siswati (Swazi), Isixhosa and Isizulu with different religious beliefs comprising of Christianity, Hinduism, Traditional African religion, Islam, Atheism, Judaism, Buddhism, Bahaism, and Agnosticism [25].

The Economy of South Africa is the second largest economy in Sub-Saharan Africa after Nigeria, with a GDP of $360 billion and a GDP per capita of $6,121 [4, 20]. South Africa is the most industrialized country in Africa, Africa's manufacturing hub and a viable market to create a $2.6 trillion GDP [28]. South Africa is the world's largest producer and exporter of gold, chrome, platinum, and manganese, the world second largest producer of palladium and the world fourth largest producers of diamond [38].

© The Author(s), under exclusive license to Springer Nature Switzerland AG 2021   119
N. Kambule and N. Nwulu, *The Deployment of Prepaid Electricity Meters in Sub-Saharan Africa*, Lecture Notes in Electrical Engineering 759,
https://doi.org/10.1007/978-3-030-71217-4_7

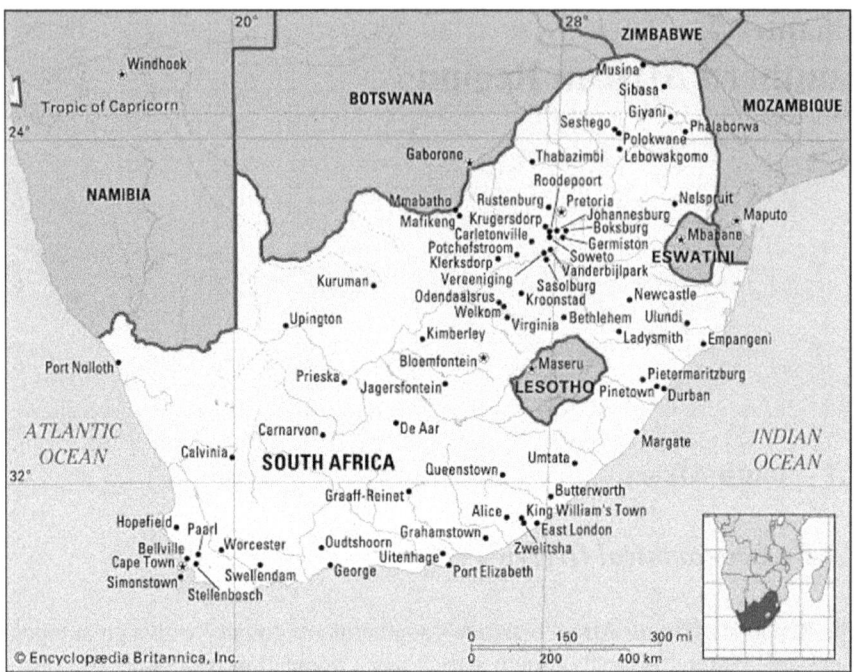

Fig. 7.1 The Republic of South Africa [5]

7.1.2   Energy Overview

Energy is the centerpiece of any development and a powerful force that powers businesses and all sectors of the economy. The energy sector of South Africa is driven majorly by coal, crude oil, renewable energy, nuclear and natural gas [42].

Narrowing down to South Africa's electrical energy, the electricity sub-sector is dominated by Eskom, which has 27 operational power plants and is the country's national utility. Eskom is majorly responsible for the generation, transmission and distribution of 96.7% of South Africa's electricity, the independent power producers (IPPs) account for the remaining. Furthermore, South Africa is the supplier of approximately 40% of Africa's electricity [21, 42].

According to GetInvest [22], electricity demand in South Africa is about 34 481 MW, where 214 487 GWh is transmitted to the end users by Eskom. Also, electricity production in South Africa as of August 2020 is 20 978 GWh [46].

In terms of electricity consumption, WorldData [50] revealed that the total consumption of electricity in South Africa is 207.10 bn kWh per year and a per capita average of 3537 kWh. South Africa also exports and imports electric energy as revealed in Table 7.1 [50].

Table 7.1 South Africa electricity distribution matrix [50]

Electricity	kWh total	kWh per capita
Consumption	207.10 bn	3 536.65
Production	234.50 bn	4 004.56
Import	10.56 bn	180.33
Export	16.55 bn	282.62

Eskom is responsible for the operation and maintenance of 95% of South Africa's transmission network in collaboration with 187 licensed municipal distribution through the Regional Electricity Distribution (REDs) at a tariff set by the National Energy Regulator of South Africa (NERSA), while the independent power producers take care of the remaining 5% of the transmission network [22, 42]. Furthermore, according to the IEA [26], electricity access in South Africa as of 2019 is 94% and approximately 3 million have no access to electricity.

The South Africa Electrical Energy sector comprises of the following stakeholders [34]:

a. Department of Energy (DoE): It is the agency responsible for planning, formulation and implementation of policy regarding energy as well as generation, transmission and distribution of electrical energy for efficiency and electrification. It also determines the generation capacity needed and how it is to be implemented by the IPPs and Eskom.
b. National Energy Regulator of South Africa (NERSA): It is an independent regulator established by South Africa's 2004 National Energy Regulatory Act, giving it power to generate licenses, enforce regulatory compliances, regulate Eskom tariff proposals and provision of National Grid codes and standards.
c. South African National Energy Development Institute (SANEDI): It is an institute helping the DoE in achieving its aims and objective by directing, monitoring and conducting necessary research and development in promoting energy efficiency, especially green energy in South Africa.
d. Eskom: This is the national utility, which is the backbone of South Africa's electricity industry as it generates, transmits and distributes electricity to various consumers. It's also the only buyer of electricity produced by various consumers' independent power producers (IPP) and ensuring electrification. It also collaborates with 137 municipalities in distributing electricity.
e. Supporting Associations: Below are associations and councils in South Africa supporting and contributing to the growth of the energy sector:

- South Africa Renewable Energy Council.
- South African Wind Energy Association (SAWEA).
- South Africa PV Industry Association (SAPVIA).
- Southern African Solar Thermal and Electricity Association (SASTELA).
- Sustainable Energy Society of South Africa (SESSA).
- South African Independent Power Producers Association (SAIPPA).
- Southern Africa Biogas industry associations (SABIA).

- Liquid Fuels and Gas Market.
- National Energy Efficiency Agency (NEEA).
- Green Cape.

Moreover, the South Africa energy sector is regulated by the following laws, policies and regulations [34]:

- National Energy Act 34/2008
- The Electricity Regulation Act (ERA) of 2006
- White Paper on Energy Policy (1998)
- White Paper on Renewable Energy (2003)
- Energy Policies for Sustainable Development in South Africa
- National Response to South Africa's Electricity Shortage
- South Africa's Renewable Energy Independent Power Producer Procurement Programme (REIPPPP)
- 2011 and updated 2016 Integrated Resource Plan (IRP) 2010–2030
- 2015 Integrated Energy Plan
- 2016 Integrated National Electrification Program (INEP)
- New Households Electrification Strategy (2013)
- 2030 National Development Plan (NDP 2013)
- Small Projects Renewable Energy Independent Power Producers (IPPs) Programme
- 2007 Biofuels Industrial Strategy
- 2012 Biofuel Mandatory Blending Regulation
- Carbon Tax Act 15 of 2019
- South Africa's Renewable Energy Policy Roadmap
- National Cleaner Production Strategy 2004.

Furthermore, according to Global Legal Insights [24] and Slabbert [44], there are plans by the South Africa National Government to unbundle the state owned Eskom into three separate state owned entities covering electricity generation, transmission and distribution. The decision was due to Eskom poor financial performance and operational performance.

7.1.3   Prepaid Electricity History and Current Status

Historically, the prepaid metering system was introduced in South Africa in 1988, when South Africa electricity supply provider (ESKOM) had a change of strategy in reaching out to those without access to electricity [23]. Eskom in 1989 engaged two manufacturers, AEG and Conlog for 10 000 meters. This action made Eskom to be the pioneers for massive deployment of prepaid meters in the World. According to Ewon [18] by the year 2000, approximately 3.2 million prepaid meters had been installed in South Africa, with the meters facilitating effective budgeting, fair sharing of energy burden, preventing arrears, preventing credit action and high reconnection costs and fostering accessibility. A prepayment system consists of prepaid meters (measurement and metering device); vending system (a system where electricity required is bought) and a revenue management system (managing the prepayment infrastructure) [23]. Jack and Smith [29] stated that prepaid electricity meters are a technological solution to non-payment of electricity fees and the challenges associated with post-paid metering system. According to the authors, prepaid metering involves the purchase of electricity before consumption with the meters displaying the number of units and having features to inform the customers when the balance is getting low. The prepaid meter has the following advantages; generation of revenue in advance of consumption, cutting down on non-payment or late payment of bills, reduces meter inspection bribery, improves cash flow and reduces administrative inefficiency [29, 33]. De Buck [11] opined that a prepaid meter is preferable due to these reasons; provides more control over amount of electricity consumed, there is no under—charging or over—charging, feature of an SMS reminders when the balance is low and topping up via cell phone, availability of usage statistics and great customer service from metering companies. In South Africa, customers can buy their prepaid electricity from electronic and/or physical vendors which includes gas stations, online/mobile phone, kiosks, ATMs, small shops and supermarkets.

In describing how the prepaid meters works, Gina [23] describe thus; the electricity consumer with a prepaid meter goes to a vending site to purchase electricity along with the meter serial number or a meter swipe card and in return, the vendor will return a token (like a paper receipt) containing 20 digits to the customer. This transaction is digitally recorded in the management system of the vendor. The 20 digits is entered into the meter and there is electricity loaded to the meter.

Moreover, Nhede [36] stated in an interview that as of 2018, there are approximately seven (7) Million prepaid meters currently installed in South Africa which could increase to eight Million by 2024.

7.1.4 Challenges of Prepaid Metering

The key prepaid metering challenge in South Africa stems from electricity theft resulting in significant loss of electricity and revenue [23, 33, 37]. The authors categorise these challenges as follows:

a. Tampering and bypassing of meters: this is a situation whereby consumers connect and hook into a power supply and distorts a meter in order to avoid recording of electricity usage. This leads to overloading and is harmful to electronic appliances. Another aspect of this challenge is tapping into neighboring premises which involves illegally connecting electric wires across fences, which could lead to fire outbreaks and electrocution of innocent people due to exposed wires and ill—concealed wires.

b.   Illegal prepaid electricity vouchers: this includes the sales and purchase of illegal
      prepaid electricity vouchers from stolen vending machine.

Furthermore, another challenge with prepaid meters in South Africa according to
Jack et al. [30], Jack and Smith [29] and Kambule et al. [31] is the exacerbation of
energy poverty. This is because it is shown that in South Africa, prepaid metering has
a negative effect on the poor who are forced to reduce their energy demand/electricity
by approximately 14% and are forced to purchase small amounts of electricity per
time, thus affecting their livelihood. Also, Nhede [36], PamL [39] and Roodbol [43],
stated that all prepaid meters worldwide, including South Africa are facing a Token
Identifier (TID) timeline challenge. According to the authors, TID is the number of
minutes that have elapsed since a defined based date of 1993 up to the time of creating
the token. The TID will run out in November 2024, where all existing prepayment
meters will stop accepting credit tokens, meaning no electricity supply. The authors
further opined that the existing meters needs to be cleared of all stored TIDs and a
change in its cryptographic key before 2024, where a new range of TID will then
start from a new base date of 2014 and run out in 2045, extending the usefulness of
the prepaid meters.

7.1.5   Future Prognosis

In line with technological advancements and the mechanical difficulties associated
with prepaid meters, Ellenki and Srikanth [16] opined that Advanced Metering Infrastructure (AMI) system in the energy sector is the game changer for prepaid energy
metering, as it integrates prepaid meters into the grid whilst simultaneously monitoring households' electric power usage. Luan et al. [32] opined that the AMI is
the communication hardware and software that creates networks between advanced
meters and utility business systems which allows collection and distribution of information to customers [1]. AMI is often synonymous with Smart Metering. According
to the [45], a smart metering system is an electricity metering system that measures
the consumption of energy on a time interval basis, allows two—way communication between the end—users and the utility, permits storage of time interval data
and remote load management. The report further stated that smart metering in South
Africa will be mandated for consumers with a monthly consumption of at least
1000 kWh. A smart meter possess the ability to store metering data in registers
and support a variety of tariffs (like Time of Use, TOU, Inclined Block Tarriff, IBT,
Maximum Demand, Free Basic Electricity (FBE), which can all be updated remotely
[17]. Gina [23] and Edison Electric Institute [15] highlighted the benefits of smart
metering for different categories and stakeholders and it is shown in Table 7.2.

Table 7.2 Benefits of smart metering [15, 23]

Stakeholder	Benefits
Municipal customers	• Efficient access and data to manage electricity use • Proactive customer notification • More accurate and timely billing • Improved outage restoration • Power quality data • Possibility of shifting consumption away from heavy demand periods and when tariffs are higher • Improved safe, reliable and stable electricity supply • Knowledge of appliances that consumes more electric energy
Customer services and field operations	• Reduced cost of meter reading • Reduced trips for off—cycle reads • Reduction in transactions and frequency call centers. • Reduction of issues surrounding connection/disconnection • Improved reliability of grid
Transmission and distribution	• Improvement in load management of transformers • Improved capacity banking switching • Improved demand management • Data for improved efficiency, reliability of service, losses and loading • Improved data efficiency grid system design
Revenue protection	• Reduction in back office rebilling • Quick and early detection of meter bypassing and meter tampering • Reduction in estimated billing and errors on bills

Furthermore, Prepaid smart meters are expected to populate South Africa cities in line with "Smart Grid Vision 2030" initiative by the South African National Energy Development Institute (SANEDI) as its readily accepted by South Africans due to it numerous benefits [17]. The prepaid smart metering market in South Africa is fast growing as it was valued at $54.1 Million in 2017 and is projected to be worth over $79.5 Million by 2025.

7.2 Mozambique

7.2.1 Geographical Overview

The Republic of Mozambique is a country in Southern African, specifically at the east coast of Southern Africa between South Africa and Tanzania at the Mozambique Channel [7]. Other countries that share borders with Mozambique include Malawi, Eswatini, Zambia and Zimbabwe as shown in Fig. 7.2, with geographic coordinates of

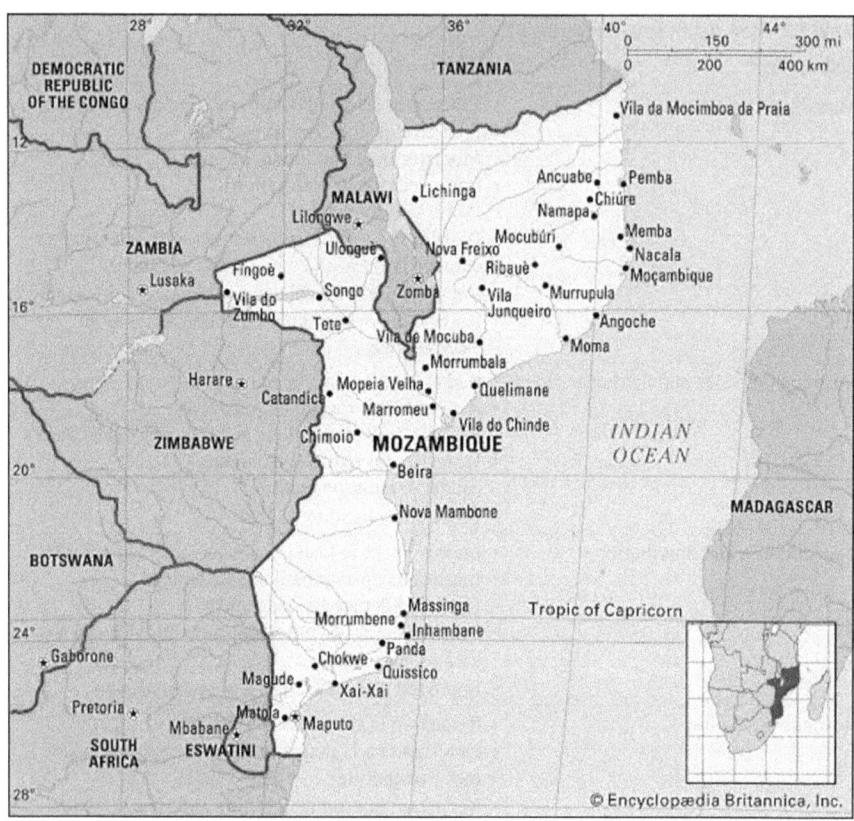

Fig. 7.2 The Republic of Mozambique [6]

18.6657°S, 35.5296°E. According to the Food and Agriculture Organization (FAO) of the United Nations [19], Mozambique's total area is 79 9380 km^2 with three geographic divisions, namely:

a. A coastal belt comprising of areas south of the Save river and the lower Zambezi area, covering 44% of Mozambique.

b. Middle Plateau area that ranges from 200–1000 m in elevation covering about 29% of the country.

c. A plateau and highland region with average elevations of around 1000 m to the north of Zambezi River covering 27% of Mozambique.

According to Worldometer [51], the population of Mozambique is estimated to be 31, 622 178 as of November, 2020 with a population density of 40 per km^2 and a life expectancy of 62.13 years. The country's capital territory is Maputo and its official languages are Makhuwa and Portuguese [8].

Economically, Mozambique is a poor country with a GDP of 15 billion USD and a GDP per capita of $483 [20]. Mozambique is a least developed country and a low—income economy with the following industries; aluminium, petroleum products, coal, chemicals, tobacco, food processing, manufacturing, agriculture and the services sector [27, 48].

7.2.2 Energy Overview

Developing countries, especially a hitherto war ravaged country like Mozambique need adequate supply of energy to promote developmental projects for the well-being of her citizens and for rebuilding the country [10]. Mozambique's energy resources comprises of hydropower, natural gas, coal, biomass and renewable energy such as wind energy, geothermic and oceanic energy [9]. According to the US International Trade Administration [47], electricity generation in Mozambique as of 2020 is 17.44 TWh. However, according to Power Africa [41], installed electricity capacity is 2827 MW with 2184 MW stemming from hydroelectric sources and 643 MW stemming from thermal sources. The derived/installed electricity generated is mostly from the Cahora Bassa Dam, where 60% of the generated electricity is utilized by the Mozal aluminium plant, 30% of the generated electricity is exported to South Africa and Zimbabwe, while the rest is used in Mozambique. Also, there are power plants that complement the generation of electricity from the Bassa Dam and they include Corumana, Chicamba, Maruzi, Cuamba, Lichinga, GTG_1, Maputo, GTG_2 Maputo, GTG_3 Maputo and GTG Beira, Termica de Temane, Gigawatt MZB, Central Termo-electrica de Ressano Garcia (CTRG), Aggreko (Nacalal) and Aggreko [21]. The generation of electricity is done by Mozambique owned utility called Electricadade de Mocambique, EdM [22].

According to the [47], electricity consumption in Mozambique is 19.6 TWh. Electrical energy in Mozambique is distributed by the Mozambique Transmission Company (MOTRACO), which supplies to the Mozal Aluminium smelter, after purchasing it back from Eskom, South Africa [35].

The NEA [35] report classified the EdM transmission system into three, namely;

a. The Northern Region: it has a 220 kV transmission system spanning 1,000 km from Songo substation to Nampula and continuing at 110 kV to Nacala town. It also has a 220 kV system but operated at 110 kV extending from Tete up to the central region of Chibata.

b. The Central Region: it has a 110 kV system linking the hydroelectric power stations in Chicamba and Mavuzi with the load centers in the corridors of Beira–Manica.

c. The southern Region: it has a 110 kV network extending from Maputo to Xaixai, Chokwe and Inhambane. It also has a 275 km single—circuit line from Maputo to Komatipoort, which connects with South Africa's Eskom.

Furthermore, according to the NEA [35], electricity consumption in Mozambique is approximately 13 449 GWh, from the following sectors industry; industry (9426 GWh), residential (1654 GWh), Commercial and Public services (702 GWh), Agriculture/forestry (30 GWh) and others (1637 GWh). However, according to the WorldData [49], there is a total consumption of electric energy per year of 11.57bn kWh and a per capita energy consumption of 381 kWh, Furthermore, only 35% of the population in Mozambique has access to electricity, with EdM planning to increase it to 49% by 2024.

The Energy [9] and NEA [35] in their reports highlighted the stakeholders in Mozambique energy sector and electricity sub-sector, which are:

a. Ministry of Energy: This government ministry is responsible for the national energy strategic planning and formulation of policy in overseeing the operation and development of the energy sector. In provinces, the ministry is represented by Provincial Directorates of Mineral Resources and Energy with goals of increasing access to electricity; ensuring the supply of reliable and quality energy and also promoting diversification of the energy matrix.

b. Electricidade de Mocambique (EdM): It is a government owned vertically—integrated electricity utility that is responsible for generating, transmitting and distributing electricity through the national grid, in a working relationship with HCB and MOTRACO. The Mozambique national grid is interconnected with South Africa, Zimbabwe and Swaziland.

c. Hidroelectrica de Cahora Bassa (HCB): this is the company that manages the generation of electricity from the Cahora Bassa Dam/Plant, the HVDC transmission system, Matambo substation and transmission lines. HCB is run as an independent power producer (IPP) but it is 92.5% owned by the government of Mozambique and the remaining 7.5% is owned by the Portuguese government (the country's former colonial master). It has a working relationship with EdM and ESKOM in South Africa.

d. Mozambique Transmission Company (MOTRACO): this is a joint venture founded in 1998 between three electricity companies in Mozambique, South Africa and Swaziland, which are Mozambique electricity (EdM), Eskom Holdings Limited (ESKOM) and Swaziland Electricity Company (SEC). MOTRACO transport electricity from Eskom to EdM and SEC.

e. Fundo Nacional de Energia (FUNAE): this is a public institution saddled with the responsibility of promoting energy access in a sustainable and rational manner through the provision and mobilization of fund and resources and machineries for energy projects across the Republic.

f. Conselho Nacional de Electricidade (CNELEC): this is Mozambique's National Electricity Council, which is an independent advisory body for the country's power sector. The council service as an advisory body, consultative body, conciliation authority, mediation authority and arbitration authority for disputes in the energy sector.

Moreover, according to NEA [35], the following are regulatory laws in the Mozambique energy sector:

- Electricity Act 1997.
- Energy Policy 1998.
- Energy Strategy 2009.
- New and Renewable Energy Development Policy (2009).
- The National Energy Strategy (2014–2023).

7.2.3 Prepaid Electricity History and Current Status

According to [12], prepaid metering system was introduced to the Mozambique electricity sector due to the following reasons; 43% total distribution losses in 1995, bad debt from domestic and public institutions, poor quality of service, high running costs associated with customers management and low income. Also, Baptisa [2] stated that prepaid metering was introduced in the country to address customer debt, eliminate electricity theft and improve the cash flow of EdM. The prepaid metering system was introduced to Mozambique in 1995 in a Pilot Project of 500 customers in Matola city [2, 12]. Also, in 1996, another pilot project was carried out in Maputo city for 5,000 customers [12]. Baptisa [3] opined that the prepayment system was compulsory for all and not targeted at poor households unlike in South Africa and UK. According to [3], Maputo, with a population of more than one million has approximately 93% of it's residents using prepaid meters as of 2012. EdM [13] stated that the prepaid system is a range of technologies that gives electricity consumers the possibility of keeping track of their costs of electricity, giving them the autonomy of deciding how much the consumer wants to spend on electricity over a given period. According to the company, prepaid metering system, which is also known as Credelec in Mozambique has the following advantages; improved management in the consumption of electricity by customers, increased transparency in processing electricity consumption, elimination of power disconnection due to delayed payment, possibility of buying electricity based on the purchasing power of the customers, no penalty due to delayed payment as experienced in post-paid era, improvement in service delivery to customers and a strengthened relationship between EdM and the customers. Baptisa [3] opined that prepayment system household in managing their disposable income and to stay out of debt by consuming only what they can afford and when they can afford.

According to the EdM [14], prepaid meters as of late 2020 have been obtained by 78% of its customers, where the electricity token can be purchased in petrol stations, vending machines in supermarkets, Automated Teller Machine (ATM), Scratch card (Voucher) vendors and via cellphone & internet banking. Baptisa [3] stated that the prepaid metering system in Mozambique has led to an increase in revenue and an upgrade in the obsolete grid left behind by the Portuguese.

7.2.4   Challenges of Prepaid Metering

In a comprehensive review of the challenges of prepaid metering system, The United States Agency for International Development (USAID) [40] stated that Mozambique's electricity utility, EdM lacks a documented metering strategy, policy, standards and roadmap that supports effective metering system in the country vis—a—vis the goals of EdM. These documented strategies and policies will provide the basis for meter procurement, meter handling, meter standards and requirements and minimization of cost associated with lifecycle of meters.

Also, Baptisa [3] analyzed the politics of the artificially low prepaid electricity tariff, where the country's electricity utility (EdM) under government pressure has been using the low electricity tariff as a political tool to curry goodwill against opposition parties, thus depriving the sector of healthy competition necessary for development.. Also, there are questions about the sovereignty of the country's energy sector due to the energy infrastructural dependency on South Africa and the influence the country's former colonial master has on the sector. Bapista [3] highlighted the following challenges of prepaid electricity:

a.   Electricity is marred by low—quality service plagued with voltage fluctuations and constant power outage.
b.   The prepaid electricity metering rate was marginally expensive than the conventional block tariff rates.
c.   The prepaid metering system retains the service fees associated with the conventional post-paid metering, which has become a point of contention with many consumers. The service fees includes; radio and television fee, collected by the central government and the refuse fee by the municipal administration.
d.   Long waiting queues due to the slowness of the sales systems.

7.2.5   Future Prognosis

The quality of electricity supply enjoyed by prepaid metering consumers is a major concern. In addition, the quality of prepaid meters is also a challenge. In order to ensure the continuous quality of prepaid meters, the Mozambique EdM has detailed plans to evaluate meter performance focusing on calibrations and reliability, which will ensure speedy meter acquisition and reduce the chances of acquiring meters with poor performance and low durability [40]. The reports detailed the processes to achieve better quality in metering stated below:

a.   Registration of meter suppliers with EdM in ascertaining the suppliers financial and logistics capabilities.
b.   Provision of meter calibration certificates in accordance with EdM metering standards.
c.   Provision of certification of meter performance under the following extreme conditions; humidity, temperature, voltage limits, vibration etc.

7.3 Highlights

- Prepaid meters were launched in South Africa and Mozambique (two Southern African nations) to serve as an arrearage recovery mechanism. Whilst in Mozambique, it was targeted at all customer classes, South Africa initially rolled prepaid meters for low-income customers.
- Electricity theft is still a serious problem amongst low-income households in South Africa whilst in Mozambique, the power quality and prepaid metering service quality still need improvement.
- Smart meters with the ability to disable itself and alert the utility upon being tampered with should be more widely deployed in both South Africa and Mozambique. Furthermore, prepaid meters should be better quality controlled, especially in Mozambique in addition to improvement of power supply quality.

References

1. Aclara (2008) AMI Industry Glossary. Y32000—REF, REV A
2. Baptisa I (2016a) Power to the People? Prepaid Electricity in Mozambique. Africa Research Institute. https://www.africaresearchinstitute.org/newsite/blog/power-to-the-people-prepaid-electricity-in-mozambique/. Accessed 17 Nov 2020
3. Baptista I (2016b) Maputo: Fluid flows of power and electricity: Prepayment as mediator of state—society relationships. In: Luque–Ayala, Silver J (eds) Energy, power and protest on the urban grid, Routledge
4. Bloomberg (2020) Nigeria's Tops south Africa as the continents biggest economy. Economics Prinesha Naidoo. https://www.google.com/amp/s/www.bloomberg.com/amp/news/articles/2020-03-03/nigeria-now-tops-south-africa-as-the-continents-s-biggest-economy. Accessed 20 Nov 2020
5. Britannica (2020a) Mozambique. https://www.britannica.com/place/mozambique. Accessed 20 Nov 2020
6. Britannica (2020b) South Africa. https://www.britannica.com/place/south-africa. Accessed 20 Nov 2020
7. CIA World Fact Book (2020a) Central Intelligence Agency, World Factbook. Library Publications, South Africa. https://www.cia.gov/library/publications/the-world-factbook/geos/sf.html. Accessed 25 Nov 2020
8. CIA World Fact Book (2020b) Central Intelligence Agency, World Factbook. Library Publications, Mozambique. https://www.cia.gov/library/publications/the-world-factbook/geos/mz.html. Accessed 25 Nov 2020
9. Charter Energy (2017) Pre-assessment report of the Mozambique energy sector under the principles of the international energy charter and the energy charter treaty. Energy Charter Secretariat, Brussels, Belgium
10. Curvilas CA, Jirjis R, Lucas C (2010) Energy situation in Mozambique: A review. Renew Sustain Energy Rev 1–11. https://doi.org/10.1016/j.rser.2010.02.002
11. De Buck X (2016) Prepaid meters in South Africa: how do they work? https://luxuryhomesjohannesburg.com/real-estate-blog/prepaid-meters-in-south-africa-how-do-they-work/. Accessed 18 Nov 2020
12. EdM (2007) Implementation of prepayment system in Mozambique EDM prepayment system. Slides. https://www.google.com/amp/s/slideplayer.com/amp/3527234/. Accessed 20 Nov 2020

13. EdM (2018) Electricidade de Mocambique (EDM) Integrated Master Plan Mozambique Power system Development Final Report. Japan International Cooperation agency (JICA). JERA Co. Inc
14. EdM (2020) Contador CREDELEC. https://www.edm.co.mz/en/products/contador-credelec. Accessed 20 Nov 2020
15. Edison Electric Institute (2014) Handbook for Electricity Metering - Eleventh Edition
16. Ellenki SK, Srikanth RG, Srikanth C (2014) An advanced Smart Energy Metering System for Developing countries. Int J Sci Res Educ 2(1):242–258
17. Eskom (2020) Eskom smart prepaid split meters programme. https://www.eskom.co.za/custom ercare/smartprepayment/pages/default.aspx. Accessed 22 Nov 2020
18. Ewon (2014) Prepayment meters. Discussion Paper. Energy & Water Ombudsman NSW. ABN 21 079 718 915
19. FAO (2016) Food and agriculture organizations of the United Nations. Country Profile–Mozambique. Version (2016) AQUASTAT. Italy, Rome
20. FocusEconomics (2020) Mozambique economic outlook. https://www.focus-economics.com/ countries/mozambique. Accessed 20 Nov 2020
21. GetInvest (2020a) Mozambique energy sector. https://www.get-invest.eu/market-information/ mozambique/energy-sector/. Accessed 20 Nov 2020
22. GetInvest (2020b) South Africa energy sector. https://www.get-invest.eu/market-information/ south-africa/energy-sector/. Accessed 20 Nov 2020
23. Gina M (2016) Customer satisfaction analysis of Conlog electricity prepayment meters in KwaZulu—Natal: a customer perspective. Master Degree Dissertation, Durban University of Technology
24. Global Legal Insights (2020) Energy 2021| South Africa. GLI. https://www.globalegalinsights. com/practice-areas/energy-laws-and-regulations/south-africa. Accessed 19 Nov 2020
25. Government of South Africa (2020) South Africa's people. Official Guide to South Africa. https://www.gov.za/about-sa/south-africas-people. Accessed 20 Nov 2020
26. IEA (2019) South Africa energy outlook. Analysis from Africa Energy Outlook 2019. https:// www.iea.org/articles/south-africa-energy-outlook. Accessed 20 Nov 2020
27. International Monetary Fund (2019) World Economic Outlook Databases. https://www.imf. org/en/publications/SPROLLS/World-economic-outlook-databases. Accessed 20 Nov 2020
28. InvestZA (2020) Why South Africa? Department of trade, industry and competition. http:// www.investsa.gov.za. Accessed 20 Nov 2020
29. Jack BK, Smith G (2016) Charging ahead: prepaid electricity metering in South Africa. Working Paper 22895. National Bureau of Economic Research. https://www.nber.org/papers/w22895. Accessed 20 Nov 2020
30. Jack K, McDermott K, Sautmann A (2019) Prepaid electricity metering and its effect on the poor. International Growth centre. https://www.theigc.org/blog/pre-paid-electricity-metering-and-its-effect-on-the-poor/. Accessed 20 Nov 2020
31. Kambule N, Yessoufou K, Nwulu N, Mbohwa C (2018) Temporal analysis of electricity consumption for prepaid metered low and high income households in Soweto, South Africa. Afr J Sci Technol Innov Dev 11(3):375–382. https://doi.org/10.1080/20421338.2018.1527983
32. Luan SW, Teng JH, Chan SY, Hwang LC (2010) Development of an automatic reliability calculation system for advance metering infrastructure. Presented at the 2010 8th IEEE international conference on industrial informatics (INDIN), Osaka, 2010
33. Maphaka M (2009) Challenges and risks in Universal provision of electricity energy in South Africa: approaches and challenges in slum classification, UN Headquarters, Nairobi, 26–28 October 2009
34. Mokverd K, Eije SV (2018) Final energy report South Africa. Netherlands Enterprise Agency. Ministry of Foreign Affairs. RVO-208-1801/RP-INT
35. Netherland Enterprise Agency, NEA (2018) Final energy report Mozambique. Ministry of Foreign Affairs. Kees Mokveld & Steven Von Eije. Version 6. RVO-207-1801/RP-INT

36. Nhede N (2018a) Every prepaid electricity user and utility should know that the token iden-
    tifier has a limited range and will run out in November 2024. STS Association interview
    on Prepaid meters in South Africa. https://www.smart-energy.com/event-news/sts-association-
    don-taylor/. Accessed 20 Nov 2020
37. Nhede N (2018b) Smart meters: ease of use and energy conservation features to aid in gaining
    popularity. Smart Energy International. https://www.smart-energy.com/industry-sectors/smart-
    meters/smart-electricity-meters-grandview-research-south-africa/. Accessed 20 Nov 2020
38. Nordea Trade (2020) The Economic Context of South Africa. South Africa: Economic and
    Political overview. https://www.nordeatrade.com/fi/explore-new-market/south-africa/econom
    ical-context#top. Accessed 20 Nov 2020
39. PamL (2018) SA's TID rollover: what every prepaid electricity user and provider should
    know. ESI Africa. https://www.esi-africa.com/industry-sectors/metering/sas-tid-prepaid-met
    ers-rollover/. Accessed 20 Nov 2020
40. Power Africa (2020) Mozambique Power Africa Fact Sheet. US Agency for International
    Development. https://www.usaid.gov/powerafrica/mozambique. Accessed 20 Nov 2020
41. Power Africa (2020) South Africa Power Africa fact sheet. energy sector overview. US Agency
    for International Development. https://www.usaid.gov/powerafrica/south-africa. Accessed 10
    Nov 2020
42. Ratshomo K, Nembahe R (2020) The South Africa energy sector report 2019. Department of
    Energy, Private Bag X96. ISBN: 978-1-920435-17-2. http://www.energy.gov.za
43. Roodbol A (2020) The clock is ticking for prepaid meter users: Here's why.
    ESI Africa. https://www.esi-africa.com/industry-sectors/metering/the-clock-is-ticking-for-pre
    paid-meter-users-heres-why. Accessed 22 Nov 2020
44. Slabbart A (2020) New CEO Andre de Ruyter to put brakes on Eskom Unbundling.
    City Press. https://www.google.com/amp/s/www.news24.com/amp/citypress/business/new-
    ceo-andre-de-ruyter-to-put-brakes-on-eskom-unbundling-20200127. Accessed 23 Nov 2020
45. Sustainable Energy Africa (2015) Smart metering: overview and considerations for South
    African Municipalities. SAMSET
46. TradingEconomics (2020) South Africa Electricity Production. Trading Economics. https://tra
    dingeconomics.com/south-africa/electricity-production. Accessed 20 Nov 2020
47. US International Trade Administration (2020) Mozambique—Energy. Mozambique
    Country Commercial Guide. https://www.privacyshield.gov/article?id=Mozambique-Energy.
    Accessed 20 Nov 2020
48. World Bank (2019) The World Bank in Mozambique. https://www.worldbank.org/en/country/
    mozambique/overview. Accessed 20 Nov 2020
49. WorldData (2020) Energy consumption in Mozambique. https://www.worlddata.info/africa/
    mozambique/energy-consumption.php. Accessed 20 Nov 2020
50. Worlddata (2020) Energy consumption in South Africa. WorldData info. https://www.worldd
    ata.info/africa/south-africa/energy-consumption.ph. Accessed 20 Nov 2020
51. Worldometer (2020) Mozambique population. https://www.worldometers.info/world-popula
    tion/mozambique-population/

Part III
Policy Implications

Chapter 8
Conclusion: Lessons Learnt and Policy Recommendations

The Fourth Industrial Revolution presents the Sub-Saharan electrical-energy landscape with an opportunity to expand sustainably—that is, improving electricity accessibility without compromising the socio-economic livelihood of vulnerable households. Thus far, the rolling out of prepaid electricity meters lack the requisite socio-economic relevance, thus rendering poor households energy impoverished. This may ultimately affect the regions ability to play its part in the global endeavour to achieve the set sustainability goals. Our final perspective maintains there is a need to revisit, reconsider, and rebuild the nature and framework of prepaid electricity meter deployment. The current model is only built to recover debt for utilities; it fundamentally falls short of considering the ill socio-economic impacts induced by the technology. So, while we maintain that the technology can be of benefit to society (i.e. both public utility and household), we believe that this can be done under certain conditions, which are mostly policy-oriented.

Below is a summated outline of ten policy recommendations, based on lessons learnt by pioneering countries that should frame the sustainable deployment of prepaid electricity meters in the Sub-Saharan region:

Socio-Economic Assessments: The programme of deployment of prepaid electricity meters has to factor in the undertaking of socio-economic assessment. The assessment will entail studying the socio-economic setting of the targeted area. This means that a series of consultations and engagements with the targeted community is necessary. The community is an integral stakeholder throughout the process of programme formulation. The goal of this exercise is to build a contextual programme that speaks to the needs of the targeted community. According to Abrahams et al. [1] "smart electricity planning considers the immediate needs of people." As the reality of the renewable energy market grows in the region, so does that of having household prosumers. Customers expect the industry to constantly engage with them and meet their needs or expectations.

© The Author(s), under exclusive license to Springer Nature Switzerland AG 2021   137
N. Kambule and N. Nwulu, *The Deployment of Prepaid Electricity Meters in Sub-Saharan Africa*, Lecture Notes in Electrical Engineering 759,
https://doi.org/10.1007/978-3-030-71217-4_8

Capacity Building: Literacy of the community is another imperative component that is to be incorporated in the deployment programme. It should not be assumed that the community understands the technology and its functionality. An extensive capacity building and awareness campaign has to be prioritised. This can be done is several ways, for example, through a planned door-to-door campaign; advertisements on different media platforms such as radio, television, and newspapers. Social media is currently and will futuristically stand the most powerful means that should be taken advantage of to raise awareness and educate.

Energy Efficiency: To ensure that the region optimally benefits from prepaid electricity meters, energy efficiency has to be another component of the programme of deployment. By proving electricity accessibility through smart prepaid electricity meters, energy efficiency has to be considered. Energy efficiency assists households to use as minimal electricity as possible, meaning that the units purchased can last longer than in energy inefficient settings. This may not matter to a high-income earner, but to an impoverished house this is important. While providing energy efficiency incentives to households, the authorities also need to devise ways of driving the energy efficiency technology prices down, by perhaps providing subsidies from *inter alia* carbon tax.

Electricity Tariffs: A special prepaid electricity meter tariff can come in handy as a way of attracting consumers to accept the technology, especially given the reality that in some parts of the region households continue to reject the technology. Moreover, lower tariffs during peak demand times can work in favour of the socio-economic conditions of poor household. Albeit Sub-Saharan Africa is a warm region, authorities can consider having seasonal prepaid electricity meter tariffs. Alternatively, the customer should be given liberty to choose from different tariff packages that are relevant to their daily lifestyle.

Incentives: The programme can include the already mentioned incentives in a form of energy efficiency and low tariffs during peak demand times. Furthermore, payment holidays can be institute whereby during weekends or holidays customers can freely use a certain amount of electricity. This postures well with socio-economic development.

Regulations: The rapid proliferation of prepaid electricity meters necessitates a formulation of specific regulations that will govern the market. Thus far no country has undertaken such a project [2]. However, it is pivotal that this is done because of the established reality of the negative effect that the technology has on low-income households. So, in order to protect poor households, regulations may be important to formulate. This is where a socio-economic assessment becomes extremely important as the regulations have to be aligned to household conditions.

Enforcement: Policies lose weight because authorities easily bypass or overlook them. We propose that there be a law that holds and ensures that policy-based regulations are carried out accordingly. This is one way of ensuring that there is the enforcement of regulations and that the households are protected even from authorities.

Monitoring: The aspect of monitoring implementation is important to ensure that the project continuously yields the desired results. To do this, a monitoring committee may be established to specifically focus on the quality assurance aspect of the project.

Ethical Leadership: Internal utility programmes geared towards training management and employees on ethical matters are important to undertake. Heavy penalties for workers involved in corrupt dealings of illegally connecting households should be carried out.

Political Buy-in: The government has to lead the public by example, meaning that they should work hand-in-hand with the utility in developing a working prepaid electricity model and encourage the public to use the technology.

Finally, the maximum results of the prepaid electricity meter programme can only materialise within a holistic approach. Proper and contextual planning by relevant stakeholders is necessary from inception. A sustainable socio-economic outcome can only be yielded under a programme that considers the foregoing aspects. In the years to come, more scholarly rigor needs to be invested in research, development, practice, as well as monitoring of the regional posture in the Fourth Industrial wave. It is a civil duty of regional authorities, and even utilities, to protect the most vulnerable and through the proposed policy recommendations this can be done effectively.

References

1. Abrahams Y, Fischer R, Martin B, McDaid L (2013) Smart electricity Planning. Project 90
2. Smart Energy International (SEI) (2017): Analysis: Prepaid electricity metering in Africa. https://www.smart-energy.com/features-analysis/analysis-prepaid-electricity-meters-africa/. Accessed 13 May 2020

Lightning Source UK Ltd.
Milton Keynes UK
UKHW021145230522
403388UK00009B/358